AGRICULTURE ISSUES AND POLICIES

MODULATION OF MEMBRANE-PROTEIN INTERACTIONS APPLIED TO WHEY FRACTIONATION

AGRICULTURE ISSUES AND POLICIES

Additional books in this series can be found on Nova's website under the Series tab.

Additional E-books in this series can be found on Nova's website under the E-books tab.

AGRICULTURE ISSUES AND POLICIES

MODULATION OF MEMBRANE-PROTEIN INTERACTIONS APPLIED TO WHEY FRACTIONATION

M. CARMEN ALMÉCIJA
R. IBÁÑEZ
A. GUADIX
AND
E.M. GUADIX

Nova Science Publishers, Inc.
New York

Copyright © 2011 by Nova Science Publishers, Inc.

All rights reserved. No part of this book may be reproduced, stored in a retrieval system or transmitted in any form or by any means: electronic, electrostatic, magnetic, tape, mechanical photocopying, recording or otherwise without the written permission of the Publisher.

For permission to use material from this book please contact us:
Telephone 631-231-7269; Fax 631-231-8175
Web Site: http://www.novapublishers.com

NOTICE TO THE READER

The Publisher has taken reasonable care in the preparation of this book, but makes no expressed or implied warranty of any kind and assumes no responsibility for any errors or omissions. No liability is assumed for incidental or consequential damages in connection with or arising out of information contained in this book. The Publisher shall not be liable for any special, consequential, or exemplary damages resulting, in whole or in part, from the readers' use of, or reliance upon, this material. Any parts of this book based on government reports are so indicated and copyright is claimed for those parts to the extent applicable to compilations of such works.

Independent verification should be sought for any data, advice or recommendations contained in this book. In addition, no responsibility is assumed by the publisher for any injury and/or damage to persons or property arising from any methods, products, instructions, ideas or otherwise contained in this publication.

This publication is designed to provide accurate and authoritative information with regard to the subject matter covered herein. It is sold with the clear understanding that the Publisher is not engaged in rendering legal or any other professional services. If legal or any other expert assistance is required, the services of a competent person should be sought. FROM A DECLARATION OF PARTICIPANTS JOINTLY ADOPTED BY A COMMITTEE OF THE AMERICAN BAR ASSOCIATION AND A COMMITTEE OF PUBLISHERS.

Additional color graphics may be available in the e-book version of this book.

LIBRARY OF CONGRESS CATALOGING-IN-PUBLICATION DATA

Modulation of membrane-protein interactions applied to whey fractionation / authors: M. Carmen Almicija ... [et al.].
    p. cm.
    Includes bibliographical references and index.
    ISBN 978-1-61209-674-2 (softcover : alk. paper) 1. Whey. 2. Whey products. I. Almicija, M. Carmen.
    SF275.W5M63 2011
    641.3'73--dc22
                                2011002562

Published by Nova Science Publishers, Inc. ✦ *New York*

CONTENTS

Preface		vii
Chapter 1	Whey Protein Properties	1
Chapter 2	Whey Protein Fractionation	17
Chapter 3	Experimental	27
Chapter 5	Results and Discussion	35
Conclusion		57
Acknowledgments		59
References		61
Index		69

PREFACE

Whey is an aqueous by-product of the cheesemaking industry which contains considerable quantities of lactose, proteins and mineral salts. This stream was considered as a waste (with an extreme pollution effect) in the past but now it is usually processed in order to obtain ingredients that find applications in the food industry, such as whey or lactose powders, whey protein concentrates and whey protein isolates. However, further value can be added if the array of proteins present in whey is fractionated since it has been demonstrated by extensive research that individual whey proteins are the responsible of a broad range of biological activities. For instance, α-lactalbumin (one of the major whey proteins) contains an important percentage of tryptophan (less abundant in other protein sources) and is suitable for infant formula and nutraceuticals. On the other side, lactoferrin, which only constitutes 2 % of the total whey protein, has strong bacterial and antiviral properties, preventing the growth of pathogenic organisms in the gut, stimulating the immune system and enhancing iron absorption. Therefore, recovery and purification of individual whey proteins is of great commercial interest for the pharmaceutical industry. Recent investigations have shown the potential of membrane technology in the fractionation of whey proteins. Most of these studies have been focused on the separation of binary model solutions of protein standards, although the fractionation of actual whey is still far from being fully accomplished by membrane filtration.

The objective of this work was to study the effect of the interactions between membrane and proteins on the separation of whey proteins. These interactions can be modulated by modifying the chemical environment of the feed solution, for example, by changing the pH. In the experimental set-up, a tubular ceramic cross-flow ultrafiltration membrane was installed and continuous diafiltration was selected as operational mode. Permeate flow was registered during the process and the concentrations for α-lactalbumin, β-lactoglobulin, bovine serum albumin (BSA), lactoferrin (LF) and immuneglobulins (IgG) were determined in both retentate and permeate side. After experiment, pH was found to be a crucial process variable since a wide variety of responses were obtained in permeate flow and protein transmission when this input was altered. As a conclusion, these results provide useful information for the design of filtration strategies in which particular whey fractions are targeted.

Chapter 1

WHEY PROTEIN PROPERTIES

Whey is a by-product of cheese and casein manufacture. It is the fraction obtained after the separation of coagulated casein from milk. This dilute liquid contains the majority of the water present in milk and all the soluble substances such as lactose, proteins, minerals, lactic acid and residual fat.

The production of whey is very high since, on average, from 9 to 12 L of whey are produced in the 1 kg cheese manufacture. Moreover, whey is a waste product with a high Biological Oxygen Demand (BOD_5), 35000-50000 mg O_2/L, one of the most polluting of the food industry, due to its great amount of organic matter content in its composition.

Until two or three decades ago, a considerable volume of whey was discharged into rivers or lakes owing to the important problems in the industrial use of this product. Nowadays, because of the fight against environmental pollution, draining whey is forbidden unless a previous waste treatment is carried out. This process has high cost and involves removing the most valuable components: lactose and proteins. Thanks to both these facts and the increasing interest on the requirements of human and animal food, currently, appropriate techniques are employed in order to use whey in industrial applications. These industrial processes are aimed at removing water, collecting mineral salts, crystallizing lactose and recovering whey proteins, without altering their properties. Therefore, the industrial applications of whey depend on the targeted component. In Figure 1, some of the principal products industrially obtained from whey are shown.

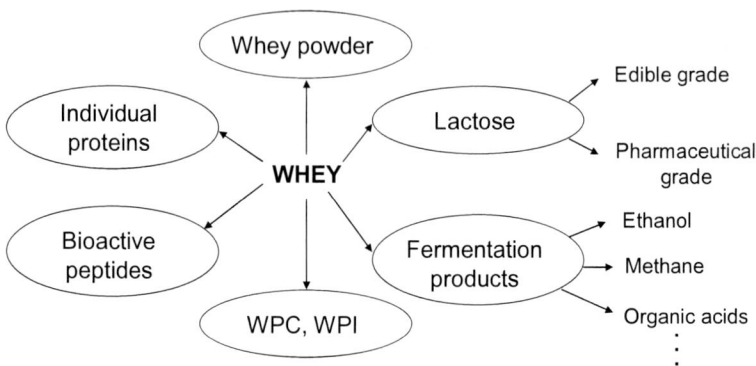

Figure 1. Major products made from whey.

The dehydration of whey is a procedure widely used in many countries due to it achieves, in a reduced volume, a product with high nutritional value which can be employed in some foodstuffs, but mainly in infant formula. Moreover, demineralised whey powder is produced using the same equipment to that used to produce milk powder, which reduces production costs. The main steps of this process are: demineralization, concentration in a multiple evaporator, partial crystallization of the lactose and, finally, the dehydration itself, usually in spray dryers.

Another industrial applications of whey is to purify lactose, which is widely used in the food and confectionary industries since it has a low sweetness (30% that of sucrose), binds flavours and aromas and increases the storage life of products. Moreover, the purest form of α-lactose is employed by the pharmaceutical industry as an excipient, being the second most used compound as filler in tablets, capsules and other oral product forms. In the purification process of lactose, the first step is a concentration of whey by evaporation. Then, the crystallization of lactose is induced and usually accelerated by adding lactose crystals. After that, the crystals are rinsed and separated from the mother liquor by centrifugation and subsequent dried in a fluidized bed dryer. To obtain a pharmaceutical grade of lactose, a refining process is necessary. It consists of redissolving the lactose crystals and treating the solution with activated carbon, which is later removed by flocculation and filtration. After crystallization, subsequent separation of the crystals by centrifugation, drying and milling, a high purity pharmaceutical grade lactose with a specific particle size distribution is obtained.

The fermentation of lactose is one of the alternatives for industrial uses of whey. Some examples are methane or biogas production by anaerobic digestion or the production of ethanol by fermentation and subsequent distillation.

Regarding the whey proteins, which represent around 20% of the bovine milk proteins, have a high biological value, i.e., they could be absorbed almost entirely by the digestive system. The main whey proteins and their concentrations (Zydney, 1998) are summarized in Table 1.

Table 1. Concentration of major whey proteins

Protein	Concentration (g/L)
β-lactoglobulin	2.70
α-lactalbumin	1.20
Immunoglobulins	0.65
Bovine serum albumin	0.40
Lactoferrin	0.10

These proteins are used extensively in the food industry owing to their wide range of chemical, physical and functional properties. Among the latter, highlight solubility, viscosity, water-holding capacity and their properties as emulsifying and foaming agents (Christiansen et al., 2004; Neirynck et al., 2004; Firebaugh and Daubert, 2005; Herceg et al., 2007).

In order to take advantage of all these whey protein properties, whey protein concentrates (WPC), which contain up to 80% protein, and whey protein isolates (WPI), which contain not less than 90% protein, could be obtained by ultrafiltration of whey. WPC manufacturing involves removing fat, increasing the temperature in order to avoid the bacterial pollution, and usually pre-concentrating (by evaporation, reverse osmosis, nanofiltration or ultrafiltration) and additional pretreatments, such as demineralization or enzymatic modifications. Finally, after the ultrafiltration step, WPC is dried, generally in spray dryers.

Concerning WPI, it could be obtain by two different methods: membrane technology or ionic exchange. In both cases, WPI is finally dried in spray dryers in order to achieve WPI powder. The main differences between both methods are that membrane technology does not alter chemically proteins and the glycomacropeptide fraction is retained with the others. Moreover, the final WPI is almost completely free of denatured proteins, provided pH is not modified and temperature is not too high.

Whey proteins, besides having their specific properties, partially hydrolized are an important source of bioactive peptides, with applications in food industry. For instance, α-lactalbumin and β-lactoglobulin are precursors of lactorphins, which have opioid activity (FitzGerald and Meisel, 1999; Pihlanto-Leppälä, 2001). These peptides have affinity for an opiate receptor located, mainly, in the central nervous system and in the gastrointestinal tract. α-lactalbumin and β-lactoglobulin are also precursors of lactokinins, which show angiotensin I-converting enzyme (ACE) inhibitory activity (FitzGerald and Meisel, 1999; Pihlanto-Leppälä, 2001). ACE has been classically associated with the regulation of peripheral blood pressure. This enzyme raises blood pressure by converting angiotensin I into the potent vasoconstrictor angiotensin II. ACE also degrades bradykinin, which is a vasodilative peptide, and stimulates the release of aldosterone in the adrenal cortex. Therefore, ACE inhibitors could exert an antihypertensive effect as a consequence of decreasing angiotensin II and a simultaneous increase in bradykinin activity. Regarding bovine serum albumin, is precursor of serorphin, which presents opioid activity (McIntosh et al., 1998); and of albutensin-A, an ACE inhibitor (Mullally et al., 1997). Lactoferricin B and lactoferrampin are biopeptides from lactoferrin. Lactoferricin B has been found to have antibacterial, antifungal, antiviral and antitumor activity, and has anti-inflammatory properties (van der Kraan et al., 2004). Concerning lactoferrampin, it presents antimicrobial activity (van der Kraan et al., 2004). Enzymatic hydrolysis is the most common method used to obtain bioactive peptides from whey. Pancreatic enzymes, mainly trypsin, are usually employed, although, other bacterial or fungal enzymes could be also utilized.

Apart from the above described properties, individual whey proteins have their own unique nutritional, functional and biological characteristics that make them very interesting for the improvement of infant formula, nutraceuticals and functional foods.

α-LACTALBUMIN

α-lactalbumin is present in milk of almost all mammalian species, with the exception of a few seals. Its biological function is the synthesis of lactose, since it forms a complex with a galactosyltransferase, which prevents this enzyme to transfer galactose to glycoproteins. α-lactalbumin is synthesized as result of the hormonal processes that induce the lactation. After synthesis, this protein is transported to Golgi apparatus, where the complex enzyme-protein occurs. This bond enhances the enzyme affinity for glucose.

α-lactalbumin is constituted by one polypeptide chain with 123 aminoacids, has a molecular weight of 14147 g/mol and is an acid protein with an isoelectric point between 4.5 and 4.8 (Zydney, 1998). Its tertiary structure (Figure 2), very compact and globular, is maintained by four disulfide bonds, and is divided into two domains: one helical (the α-domain) and the other α-sheet (the β-domain).

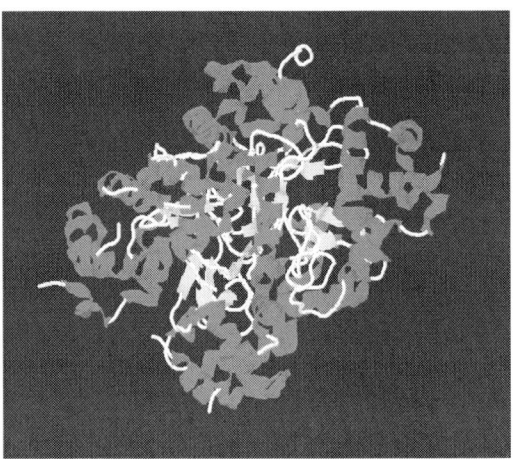

Figure 2. Diagram of the structure of α-lactalbumin.

α-lactalbumin is the second protein in concentration in bovine whey, and the most abundant in human whey. It has a calcium ion bound, which is essential to maintain its structure and its activity as a regulator of the galactosyltransferase. The removal of calcium produces the structure called "molten globule", which is an intermediate folding state, characterized by a conserved secondary structure but fluctuating tertiary structure.

α-lactalbumin, in solution, forms aggregates with different number of molecules as a function of pH (Yang et al., 2006). For instance, at pH=3, the "molten globule" state occurs. From pH=3 to pH=1.5, this structure is maintained, but its tertiary structure tends to be more flexible as pH decreases. Between pH=4 and 4.5, this protein is found as dimer. No aggregation occurs in the range of pH between 6 and 8.5. Above pH=9.5, expansion takes place but without aggregation (Boye et al., 1997).

A major application of α-lactalbumin is its use as a nutraceutical because of its high content in tryptophan (Beulens et al., 2004). Brain serotonin or 5-hydroxytryptamine is an important regulator of appetite, macronutrient preference and mood (Beulens et al., 2004). An increase in serotonin is thought to be associated with a decrease in appetite, an increased relative preference for protein instead of carbohydrate and an improvement in mood. Serotonin is synthesised from the amino acid tryptophan, which competes for transport across the blood brain barrier with the other large neutral amino acids: valine, leucine, isoleucine, tyrosine and phenylalanine. Consequently, the uptake of tryptophan depends on the total concentration of this amino acid in plasma, but mainly on the plasma ratio between tryptophan and the sum of the other large neutral amino acids. Therefore, an increase in this ratio could increase the uptake of tryptophan in brain and, consequently, an increase in serotonin levels. Tryptophan is the limiting amino acid in most protein sources; indeed, most of them decrease the ratio tryptophan-other large neutral amino acids in plasma. However, α-lactalbumin is a potential candidate for tryptophan supplementation due to it could increase this ratio up to 48% as compared to casein (Markus *et al.*, 2000).

Another α-lactalbumin application is as an additive in infant formula (Lien, 2003). Human and bovine milk vary considerably in the ratio of whey to casein protein (≈ 60:40 in human milk and ≈ 20:80 in bovine milk) and in the quantities of specific proteins. Therefore, if WPC were added in infant formula: (a) to maintain the ratio whey protein-casein present in human milk, an excess of β-lactoglobulin and deficit in α-lactalbumin would occur; (b) and to maintain the amino acid profile of human milk, an excess of protein would exist. In order to achieve infant formulas similar to human milk, α-lactalbumin is being added, involving the advantage of tryptophan concentrations in plasma equal to those in breast milk and removing the disadvantage of preceding formulas in which a not good excess of nitrogen was provided to the newborn.

Recently, a folding variant of α-lactalbumin was discovered, which selectively enters tumour cells and induces an apoptosis like mechanism (Chatterton *et al.*, 2006). It consists of the molten globule state, which is stabilised by a fatty acid cofactor. These protein-lipid complexes are called HAMLET/BAMLET for human/bovine α-lactalbumin, respectively. It is noteworthy that this interaction is stereo-specific and only unsaturated cis-fatty acids bind to α-lactalbumin and only the oleic acid bound to α-lactalbumin in a compact conformation is active against tumour cells. However, although these newly described α-lactalbumin compounds could be considered potential candidates for therapeutic treatments, further scientific and/or clinical investigation is required in order to determine if such complexes are or not formed at any stage during digestion and the health benefits for human digestion (in particular in neonates).

β-LACTOGLOBULIN

β-lactoglobulin is the major protein of bovine whey, representing around 50% of total whey protein. Although β-lactoglobulin could bind in vitro to a great variety of hydrophobic substances, including retinol and long-chain fatty acids, its physiological function is not clear. Pérez *et al.* (1992) demonstrated that β-lactoglobulin could participate in the digestion of milk lipids during the neonatal period. On the other hand, Kushibiki *et al.* (2001) studied the effect of β-lactoglobulin on the intestinal absorption of retinol, triglyceride and long chain fatty acids in pre-ruminant calves, and a beneficial effect was observed. Therefore, it was speculated that β-lactoglobulin could play a role in the absorption and subsequent metabolism of fatty acids.

This protein is formed by a 162 amino acids chain with a molecular weight of 18362 g/mol and isoelectric point of 5.2 (Zydney, 1998). In Figure 3, the structure of β-lactoglobulin is shown. There are different genetic variants, but the most common are the A and B variants, which differ in two amino acids. The A variant has valine in position 118 and aspartic acid in position 64, whereas the B variant has alanine and glycine, respectively.

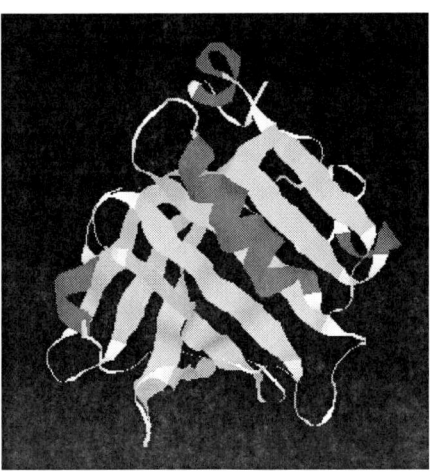

Figure 3. Diagram of the structure of β-lactoglobulin.

The tertiary structure of the β-lactoglobulin monomer has a thiol group (-SH) and is strongly stabilized by two disulfide bonds (-S-S-), which seem to play an important role in the reversibility of β-lactoglobulin denaturation. Although is the most hydrophobic of the whey proteins, β-lactoglobulin is very water-soluble due to the internal position of most its apolar components and the superficial position of the polar ones.

Depending on solution conditions, β-lactoglobulin can exist in one of its different pH dependent structural states (Taulier and Chalikian, 2001). Below pH=3, β-lactoglobulin is present as a monomer. Around pH=4.5, it tends to form aggregates up to 8 molecules. At milk pH, it is present as a dimer with the monomers bound non-covalently. These dimmers are formed between pH=4 and pH=8. Above pH=8, irreversible denaturation of β-lactoglobulin occurs and it corresponds to the monomer which loses its tertiary structure and part of its secondary structure.

Due to its excellent gelation properties, ingredients enriched in β-lactoglobulin find application in areas where water binding and texturisation are required (Chatterton et al., 2006). For instance, it is used in meats, reformed fish products and a variety of formulated foods. Moreover, the range of application of this protein could be expanded thanks to the flexibility in gel formation, since these gels can be translucent or opaque, and elastic or inelastic depending on the chemical conditions, mainly pH and ionic strength, during gelation (Chatterton et al., 2006).

Another β-lactoglobulin application is as foaming and emulsifying because of its surface active properties (Dunlap and Côté, 2005). In this sense, it is widely used in confectionery products such as meringues.

β-lactoglobulin shows high solubility and clarity over a broad range of pH, and is stable to high temperature treatment under these conditions. In addition, it has a high nutritional value. Consequently, β-lactoglobulin is a suitable active agent in various protein enriched beverages, such as fruit juices and sports drinks, and in beverages with long shelf life (Chatterton et al., 2006).

BOVINE SERUM ALBUMIN

Bovine serum albumin (BSA) is one of the most abundant proteins in blood circulatory system. Its main biological function is owed to its ability to bind ligands (Grybos et al., 2004). In this way, BSA transports fatty acids, which are water insoluble; isolates free oxygen radicals or inactivates lipophilic metabolites that are toxic.

BSA is a large globular protein, 69000 g/mol (Zydney, 1998), formed by a 583 amino acid chain. It is an acid protein like α-lactalbumin and β-lactoglobulin, with an isoelectric point of 4.7-4.9 (Zydney, 1998). Its three-dimensional structure (Figure 4) includes tree domains, each composed of ten helical segments. Its secondary structure, in its native form, is mostly alpha-helix and contains 17 disulfide bonds.

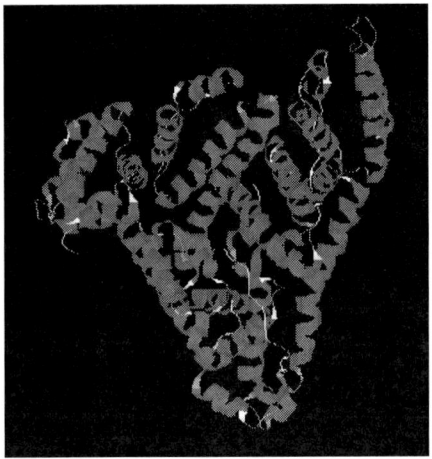

Figure 4. Diagram of the structure of bovine serum albumin.

Bovine serum albumin undergoes reversible conformational isomerisation with changes in pH. In Figure 5, the different isomeric forms and the transition pH between them are shown (Foster, 1977).

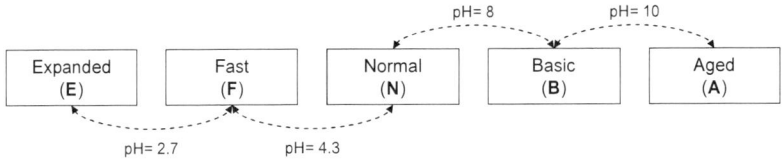

Figure 5. Isomeric forms of bovine serum albumin as a function of pH.

The N-F transition involves the unfolding of domain III. The F form is characterized by a great increase in viscosity, much lower solubility and a significant loss in helical content. Below pH=4, BSA suffers another expansion, due to the loss of the intra-domain helices (helix 10 of domain I which is connected to helix 1 of domain II; and helix 10 of domain II connected to helix 1 of domain III). This expanded form is known as the E form and is characterized by a rise in intrinsic viscosity and an increase in the hydrodynamic axial ratio. However, recent research works (Grybos et al., 2004) concluded the contrary. These authors obtained real diameters of the BSA molecule of 10 ± 2.5 nm, 11 ± 4 nm, 13 ± 6 nm y 22 ± 5 nm for pH=2.0, 3.5, 5.0 and 7.0, respectively. At pH=9, BSA changes its conformation to the basic form (B). The A form appears when solutions of BSA are maintained at pH=9 and low ionic strength for 3 or 4 days.

Bovine serum albumin is widely used both in the food industry and for therapeutic treatments (Zydney, 1998). These applications are owed to its major properties: foaming capability, gelation properties and its ability to bind ligands. The first is a consequence of its diffusion through the air-water interface, which allows it to reduce surface tension, creating and stabilizing the emulsions. At the interface, BSA molecules partially unfold and associate to produce an intermolecular film with some degree of elasticity. Its foaming capability is enhanced when BSA interacts with other proteins, mainly basic, such as protamins (Glaser et al., 2007), lysozyme and clupeine (Poole et al., 1984). When BSA is used alone, it performs better in pH values around to its isoelectric point when electrostatic repulsion is at its minimum. However, when BSA is used linked to basic proteins, the greatest expansion and stability of the emulsion is found between pH=8 and 9, which are intermediate between the isoelectric point of BSA and basic proteins, i.e., when proteins are oppositely charged.

At room temperature, the tertiary structure of BSA is well defined and stabilized. However, when temperature increases, some molecular regions loss their conformation, making them accessible to new interactions with other molecules, which produces soluble aggregates through disulfide and non-covalent bonds (Militello et al., 2003). Gelation mechanism occurs in two steps. In the initial step, an unfolding or dissociation of the protein molecules takes place, whereas association or aggregation reactions occur in the second step. The nature of gels obtained (ordered or disordered, structure and type of aggregates) depends on the conditions of the gelation process, mainly pH, temperature, heating time and protein concentration (Militello et al., 2004).

Probably, the most outstanding property of BSA is its ability to bind reversibly a great variety of ligands. Among them, it is important to highlight the long-chain fatty acids, insoluble in plasma, which are transported through its link with BSA (Álvarez et al., 1996). These fatty acids bound to BSA are not toxic and are employed to form different tissues in the body. However, in their free form, they have detergent properties, denature proteins and could damage cell membranes. Moreover, BSA transports and is the reservoir of nitric oxide, which has been implicated in important physiological processes such as neurotransmission (Stamler *et al.*, 1992). It also plays a significant role as an antioxidant (Kouoh *et al.*, 1999) due to sequestering free oxygen radicals, removed, therefore, from blood plasma.

IMMUNOGLOBULINS

Immunoglobulins are glycoproteins whose essential function is as antibodies. Immunoglobulins bind specifically to one or a few antigens; indeed, each immunoglobulin binds to a specific antigenic determinant. For the antigen destruction, collaboration of other elements is required. Thus, when immunoglobulins detect antigens, bind them and act as information transducers of their presence in order to those would be destroyed by macrophages, polymorphonuclear or NK cells.

Immunoglobulins can be classified into five different classes, based on differences in the amino acid sequences in the constant region of the heavy chains. These classes are: IgM, IgA, IgG, IgD and IgE. The immunoglobulin G (IgG) is the most abundant, representing over 70% of serum immunoglobulins, whereas IgM and IgA represent 5-10% and 10-15%, respectively. Regarding IgE and IgD, they are presented in a much smaller proportion.

Immunoglobulins have an isoelectric point between 5.5 and 8.3 and a molecular weight in the range from 150000 to 1000000 g/mol (Zydney, 1998). Although each type of immunoglobulins can differ structurally from the others, they all are built from the same basic units. Each basic structural unit presents Y-shape (Figure 6) and is composed of two identical low molecular weight chains, called light or L chains; and two identical high molecular weight chains, called heavy or H chains. The light chains have around 200 amino acids, whereas the heavy ones have around 400 amino acids. Each light chain is bound to a heavy chain by a disulfide bond, while the heavy chains are held together by two disulfide bonds.

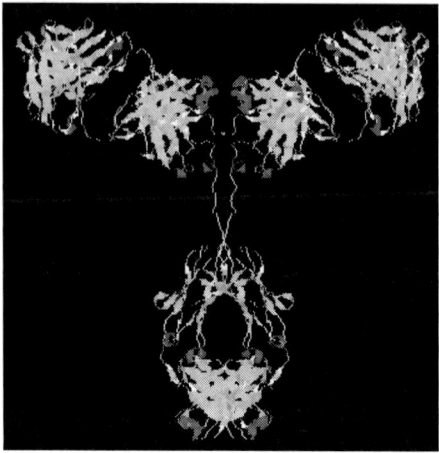

Figure 6. Diagram of the structure of a basic unit of immunoglobulins.

Mehra *et al.* (2006) summarised the major properties of immunoglobulins in their review. Among them, immunoglobulins can prevent the attachment of pathogen to the epithelial lining, which is a critical step in the establishment of infection. Consequently, immunoglobulins confer passive immunity against microbial infections. On the other hand, in the treatment of already established infections these proteins could be used with therapeutic effects, but it has been reported only in such diseases where the infection is maintained through a constant reattachment and reinfection, e.g. inside the oral cavity, and where toxins or inflammatory compounds are involved. Therefore, a few commercial preparations enriched in IgG targeted at health promotion in farm animals and humans are currently on market and more applications can be expected in the coming years.

LACTOFERRIN

Lactoferrin (LF) is an iron fixer protein, structurally related to blood transferrin and egg ovotransferrin. Milk lactoferrin has low iron saturation, since one of its biological functions is newborn protection through iron absorption, making it unavailable to bacteria and for the formation of free radicals in oxidation reactions.

Lactoferrin is a glycoprotein with a molecular weight of 78000 g/mol and a basic isoelectric point, 9.0 (Zydney, 1998). The aminoacid sequence of bovine lactoferrin is a simple polypeptide chain of 689 aminoacids (Moore et al., 1997), two aminoacids less than human lactoferrin. Despite the fact that the sequences of human and bovine lactoferrin match by 69%, both molecules have the same three-dimensional structure. Figure 7 shows that lactoferrin has two lobes (N lobe and C lobe, which corresponds to the half of the molecule containing the N terminal and C terminal, respectively), which have a 40% matching sequence approximately. These lobes, which are divided in two domains (N1, N2, C1 and C2) (Baker and Baker, 2005), are joined by a three-turn helix.

Iron unions are located in equivalent positions in both lobes. Iron-binding is reversible and takes place in the presence of a carbonate or a bicarbonate ion per ferric ion. Iron is linked directly to lateral groups of two thyrosines, and the carbonate ion (or bicarbonate ion), also attached to iron, interacts with the side chain of an arginine.

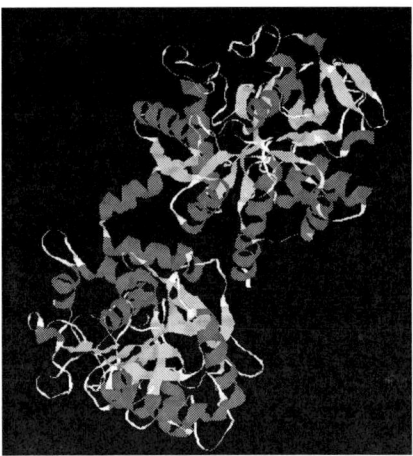

Figure 7. Diagram of the structure of lactoferrin.

Bovine lactoferrin forms non-covalent complexes with β-lactoglobulin and the BSA, with a molar-relation lactoferrin-protein of 2:1 and 1:1, respectively (Lampreave et al., 1990). However, there is no evidence about the association between lactoferrin and α-lactalbumin.

Due to its similarity to transferrins, first research works about functions of lactoferrin were focused on its ability to link iron: iron absorption, antimicrobial activity and regulation of iron metabolism during inflammation processes. However, posterior investigations have showed many other functions, some of them unrelated to iron-binding. Figure 8 shows these potential functions.

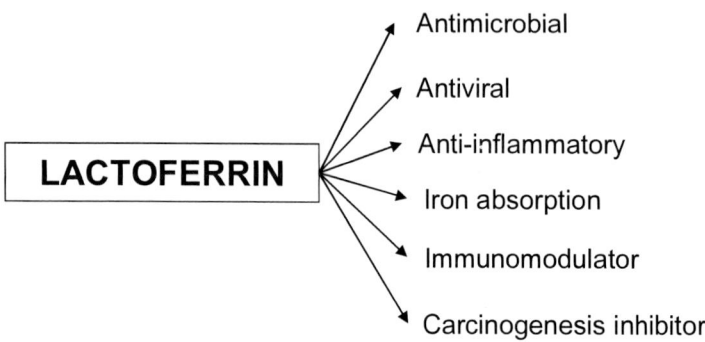

Figure 8. Main functional properties of lactoferrin.

One of the most important functions of lactoferrin is its antimicrobial activity against a great quantity of bacteria, fungi and viruses. Its first antimicrobial function is bacteriostatic, i.e., it stops bacterial growth. Due to its ability for iron-binding, lactoferrin withdraws it from infection areas so, pathogens can not use this essential nutrient for their growth. In addition, lactoferrin has a second antimicrobial function, through direct action on microorganisms: apolactoferrin (lactoferrin form without iron) can bind to the bacterial outer membrane, causing a fast release of lipopolysaccharides which increases membrane permeability, and makes it more susceptible to osmotic shock, to lysozyme and/or other bacterial molecules. Recently, a third antibacterial function of lactoferrin has been discovered: proteolytic activity. By proteolysis, lactoferrin degrades some microorganisms, so it weakens their pathogenicity (Valenti and Antonini, 2005). Regarding to its antiviral activity, bovine lactoferrin is a strong inhibitor of the development of different viruses such as herpes simplex virus 1 and 2, human immunodeficiency virus or hepatitis B and C viruses (Valenti and Antonini, 2005)

Another lactoferrin important function is the regulation of anti-inflammatory and immune responses. At cellular level, lactoferrin acts as a promoter of activation, differentiation and proliferation of immune system cells (Legrand et al., 2006). At inflammation points, lactoferrin removes free iron molecules and therefore, this iron can not serve as catalyst in the production of harmful free radicals (Legrand et al., 2005). In addition, lactoferrin can accomplish this anti-inflammatory function directly because it can bind bacterial endotoxins (lipopolysaccharides), mediators in the inflammatory response in bacterial infections (Brock, 2002).

In recent years, some studies have been done in animals about the inhibitory effect of lactoferrin in the development of experimental tumours when it is provided orally in the post-initiation state. It has been found inhibition in different cancers such as colon, esophagus, lung and bladder in rats (Tsuda et al., 2002). Inhibition in colon cancer seems to be due to the suppression of phase I enzymes, such as cytochrome P450 1A2 (CYP1A2), which is induced by carcinogenic heterocyclic amines. Another possible mechanism which has been found in metastatic cells of the small intestine involves stimulation of Interleukin 18 (IL-18) and caspase-1 production in epithelial cells, which leads to a positive induction of interferon cells (INF-γ+). This implies that the immunity of the intestinal mucosa improves substantially. On the other hand, lactoferrin exerts an anti-virus activity in patients with chronic active hepatitis C. Finally, by improving microflora, lactoferrin inhibits the production of bile free acids (toxic) which promote cell proliferation and, therefore, carcinogenesis. So, as conclusion, by oral ingesta of bovine lactoferrin, colon cancer can be prevented, metastasis of tumour cells can be reduced, intestinal immunity can be increased and it acts as an anti-virus agent of hepatitis C.

As summary, whey proteins, individually, have biological, nutritional and functional properties that make them very interesting for the improvement of infant formula, functional foods and nutraceuticals. This is the reason why whey fractionation for recovery and isolation of its proteins has an enormous scientific and commercial interest.

Chapter 2

WHEY PROTEIN FRACTIONATION

Methods used for the isolation of whey proteins explode differences between these proteins: solubility, size, concentration, charge or affinity for other molecules. In this way, the main processes proposed for the fractionation of whey proteins can be classified into three groups:

- Chromatographic methods.
- Selective precipitation.
- Filtration by membrane technology.

CHROMATOGRAPHIC METHODS

Chromatographic methods consist in a mobile phase which flows through a stationary phase. The stationary phase can adsorb the components of the solution in a selective way. Component velocity across the stationary phase is inversely proportional to its interaction with this phase. Four chromatographic methods can be distinguished according to the used medium and the elution method (Mathews *et al.*, 2000): ion exchange chromatography, affinity chromatography, gel filtration chromatography and high performance liquid chromatography.

Ion exchange chromatography is used for molecule fractionation taking into account their electric charge by using polyanionic or polycationic resins. Proteins with same sign charge as resin are not adsorbed on it and can be found in the first collected fractions. However, opposed charged proteins are linked on the resin with a strength which is proportional to their charge.

Then, a new mobile phase is used for the deactivation of proteins-resin unions which allows the recovery of the linked proteins.

Affinity chromatography is very specific and it is ideal for the isolation of one or a few proteins of a complex mixture. Some proteins show important interactions with other molecules, therefore, by using the right molecules covalently associated on a matrix of inert material, we can catch the target protein while the rest of the proteins pass through the column without any kind of interaction. Then, caught protein can be released by employing a buffer containing free matrix molecules or other reactive which can remove the interaction.

In the case of gel filtration, the separation principle is molecular size instead of chemical properties. The column is filled with spheres of a porous gel that allows the smaller molecules enter in the pores while the larger ones do not. Therefore, the larger molecules will move more quickly through the column because they can only flow through the interstices between the gel spheres and are collected in the early fractions.

Chromatography is a slow process. In normal processes, small pressures are applied in order to force fluids to pass through the column, so elution takes a long time. This process is time consuming and can also cause the deterioration of fragile materials. Additionally, the sample tends to spread by diffusion during its way through the column which affects resolution. High-performance liquid chromatography has been developed for solving all these problems. In this technique, the solution moves quickly through the column thanks to the application of a high pressure. In this way, separations, that previously required hours, can be made in only a few minutes and with better results.

The majority of the works done during last years have been focused on the utilization of these techniques for the recovery of a single protein. Thus, Schlatterer et al. (2004) isolated β-lactoglobulin from acid whey by using ceramic hydroxyapatite chromatography followed by size exclusion chromatography, obtaining a β-Lactoglobulin enriched fraction with a purity of 99%. Conrado et al. (2005) used a high-density hydrophobic resin for the recovering of α-lactalbumin from bovine whey; the product purity was 79%. Pessela et al. (2006) proposed a process for the purification of IgG from a whey protein concentrate by employing two stages of selective adsorption; they recovered an 80% of immunoglobulins with a high purity. Wolman et al. (2007) studied LF obtention from bovine colostrum whey by using a hollow fiber membrane with an affinity ligand bound. This process resulted in a final product with 94% LF purity. However, some works have studied simultaneous recovery of several proteins, Neyestani et al. (2003) used gel filtration and ion exchange in order to fractionate α-lactalbumin, β-Lactoglobulin and BSA from bovine acid whey obtaining around 170 mg of β-Lactoglobulin, 54.5mg of α-lactalbumin and 11.5 mg of BSA from 50 mL of bovine milk.

SELECTIVE PRECIPITATION

It is one of the simplest and oldest methods used in fractionation of protein mixtures. Structural stability of a globular protein depends on a series of well-balanced chemical interactions which define its final conformation. The conformation of a functional protein is partially stable but any change in the environment which surrounds it can provoke changes in its structure and a loss of his function; when this takes place, it is said that protein suffers denaturalization. Denatured proteins can be aggregated easily, and according to the environmental conditions, they can precipitate. Therefore, selective precipitation is based on creating conditions where the target protein is not denatured while the other components are denatured and can precipitate. This technique can use changes in temperature, ionic strength, pH or organic solvents in order to achieve this objective. Although the principle of these changes is different, they are not independent, for example, denaturalization by temperature is heavily dependent of pH and vice versa, and in the case of fractionation using organic solvents, pH and ionic strength must be carefully selected.

Several research works about the application of this technique to the fractionation of whey or purification of its proteins have been done during last years. In 2005, Fuda *et al.* optimized the selective separation of β-lactoglobulin from sweet whey by using stabilized microbubbles which were generated by intense agitation of a cationic surfactant. Using this process, 80-90 % β-lactoglobulin was withdrawn from the liquid phase as a precipitate. With the same technique but with an anion surfactant, Fuda *et al.* (2004) separated 90% of LF and lactoperoxidase from sweet whey. However, process was low due to the adsorption of other proteins. Ekici *et al.* (2005) fractionated BSA, β-lactoglobulin and α-lactalbumin from whey by forming foams; after the application of a two-stage process, they recovered pure BSA, 88% of β-lactoglobulin and 12% de α-lactalbumin. In 2006, Lucena *et al.* precipitated simultaneously α-lactalbumin, BSA and immunoglobulins from a whey protein concentrate, getting a enriched fraction of α-lactalbumin (recovery of 86% with a purity of 74%) and another fraction enriched in β-Lactoglobulin (99 % of the initial protein with a purity over 85%). Finally, Casal *et al.* (2006) developed a method for the selective precipitation of β-Lactoglobulin from sweet whey which allowed 100% β-Lactoglobulin recovery, while 80% of the rest of the proteins remained soluble.

FILTRATION BY MEMBRANE TECHNOLOGY

Although protein purification based on chromatographic methods is excellent for the production of protein isolates with a high purity, the scale up of these processes is difficult ought to the small quantity of protein obtained and the extremely high cost of the equipment (Ghosh and Cui, 2000a). Selective precipitation has low cost but its outputs are not adequate. Therefore, chromatographic methods and selective precipitation have not been established at commercial scale due to a bad ratio between yield/purity and a high economic cost (Cheang and Zydney, 2003). However, membrane tangential flow filtration is an attractive option with the following benefits (Ghosh and Cui, 2000a; Prádanos *et al.*, 1996):

- Mild operating conditions are required.
- It allows an efficient purification at low temperatures, so it can be used with heat-labile substances. Furthermore, the products retain their functionality because they do not suffer chemical denaturation.
- The product is free of contaminants added during the fractionation process.
- It is an easily scale-up technology

However, tangential flow filtration is a technique that is limited to the separation of solutes with a significant difference in size (Cherkasov and Polotsky, 1996), which restricts its application to the separation of proteins-cells mixtures, protein clarification or protein recovery in the biotechnology industry (van Reis et al., 1997). The constraints which hinder its application in fractionation are diverse, including concentration polarization which reduces selectivity, membrane fouling or protein-protein interactions

During the 90s, an important research work was done in order to overcome these constraints and use tangential flow filtration in fractionation of solutes with similar molecular size. The result of that effort is the development of a new technique of tangential flow filtration known as High

Performance Tangential Flow Filtration. This technique can be defined as a unitary operation which explodes both steric mechanisms and electrostatic interactions through a proper selection of physical-chemical environment. High performance tangential flow filtration can make purification, concentration and ion exchange simultaneously, allowing the combination of these three separation steps in one easily scale-up operation with an additional cost reduction (van Reis and Zydney, 2001).

The following premises must be followed for using this new technique in protein fractionation:

- Transmembrane pressure must be in the zone where filtrate flow is pressure-dependent in order to increase resolution (van Reis et al., 1997).
- Protein fractionation is optimized by handling physical-chemical properties of both membrane and proteins (Nyström et al., 1998). Membrane charge, hydrodynamic volume and protein diffusion coefficient are modified through changes in pH, ionic strength and buffers (Saksena and Zydney, 1994). This alters both protein-membrane interactions and protein-protein interactions (Howell et al., 1999).
- Selectivity can be improved by controlling the surface properties of the membrane in order to modify the exclusion of specific species through a modification in the interactions between the proteins and the membrane.
- Diafiltration mode is used (van Reis and Zydney, 2001). In this mode, the impurity (or the product) is washed from the retentate and buffer solution is added at the same time with a flow which is equal to the permeate flow. With buffer addition, protein concentration is maintained, fouling is minimized and aggregation/denaturation processes are decreased.

In recent years, there have been numerous studies focused on protein separation employing this technique. These studies have shown encouraging results which present this process as the most powerful from a technical point of view and the most feasible regarding to operational costs for being successfully implemented on an industrial scale. The research works in this field can be grouped in three groups: single model protein solutions, model protein mixtures and real mixtures.

The majority of the works focused in the filtration of single model proteins are designed to investigate electrostatic interactions between the membrane and the protein. In order to do that, these works study the effects of the most important operational parameters, mainly pH and ionic strength, over permeate flow and protein transmission. A summary of the most recent research works about the filtration of single whey protein solutions is shown: Marshall et al. (2003) tested the effect of both ionic strength and the presence of calcium ions on the filtration of β-lactoglobulin with a zirconium oxide membrane. Results showed that an increase on ionic strength decreased fouling resistance and increased protein transmission. In 2005, Rao and Zydney evaluated the possibility of controlling BSA transmission by handling both protein and membrane charge. They used a small and highly charged ligand which that selectively binds to the protein of interest. The addition of the ligand decreased BSA transmission in two orders of magnitude when a negative charged membrane was employed, however this effect was eliminated with high ionic strength or when a neutral membrane was used. Zulkali et al. (2005) studied the effect of pH, ionic strength and pressure in the filtration of BSA with an organic membrane. The fluxes were higher at lower ionic strength and at higher pH due to lower shielding of charge and electrostatic repulsion between protein molecules with similar charge. They also observed that a pressure increase provoked a flux reduction by increasing the density of concentration polarization and also stimulated rejection. In 2007, De la Casa et al. studied the influence of pH and ionic strength in tangential flow filtration of BSA with a tubular ceramic membrane. It was observed that at the isoelectric point of the protein, the permeate flow was minimum and protein transmission was maximal. Besides, a significant protein transmission was observed at the point of zero charge of the membrane. Salt addition enhanced both transmission and permeate flow. In the same year, Ibáñez et al. (2007) studied the electrostatic interactions between protein and membrane in the ultrafiltration of two model proteins: β-lactoglobulin y lysozyme, with acid and basic isoelectric points respectively throuhg a tubular ceramic membrane. pH modification showed that the lower permeate flux and the higher transmission were obtained at the isoelectric point of each protein. However, while for β-lactoglobulin a transmission decrease with operation time was obtained, lysozyme transmission remained virtually constant during the experiment. Salt addition improved the initial flux of both proteins but had a different effect over protein transmission; it provoked an increase on β-lactoglobulin transmission and a decrease on lysozyme transmission.

Studies about HPTFF applied for protein fractionation in model mixtures analyze protein-protein interactions during the filtration process. These research works search the optimization of the factors which control fractionation like pH, ionic strength, membrane cut-off, transmembrane pressure, tangential velocity, operation mode or temperature in order to obtain a good ratio purity/yield. The most relevant works in this field during last decade is now presented:

In 1997, Van Reis *et al.* fractionated a BSA-IgG mixture by analyzing the optimal conditions of pH and ionic strength. Continuous diafiltration mode with an organic membrane was used, after 80 diavolumes, a IgG fraction with a purity factor of 30 and a yield of 84% was obtained. Nyström *et al.* (1998) studied BSA-lactoferrin fractionation by using regenerated cellulose membranes. The pH employed during the filtration was equal to the lactoferrin isoelectric point, in these conditions, BSA and membrane had a high negative charge which hindered the transmission of this protein. It was observed that an increase on the total protein concentration but maintaining a 50% ratio for each one, improved the retention values for both proteins and also improved the selectivity for LF. When the mixture had more LF than BSA, the BSA selectivity was stimulated. Transmission of both proteins became better when salt was added. Finally, the effect of pH and pressure in the fractionation was investigated, the best results were obtained at pH 8.5 and 0.2 bar, with 100% BSA retention. Ghosh and Cui (2000b) made a theoretic study about the effect of tangential velocity and permeate flux on selectivity in the tangential flow filtration of a binary protein mixture. Results showed that, for constant tangential velocity, selectivity was enhanced with an increase on permeate flux up to reach an optimum; further increase on permeate flux had a negative effect on selectivity. The value of this optimal flux was function of the tangential velocity used. However, for very low or very high values of permeate flux, tangential velocity had no effect on selectivity; for a range of intermediate values, an increase in velocity caused an increase in selectivity. Cheang and Zydney (2003) analyzed the effect of pH, ionic strength, velocity through pores and membrane material on the separation of a α-lactalbumin/β-lactoglobulin mixture by ultrafiltration. After 16 diavolumes, the purification factor for β-lactoglobulin was higher than 100 with a 90% protein recovery. In the permeate, the recovery of α-lactalbumin was greater than 95%, with a purification factor bigger than 10. In 2004, Chan *et al.* used ultrafiltration for the fractionation of binary mixtures of BSA and β-lactoglobulin to indentify the apparent critical flux and to study the mechanisms and factor affecting membrane fouling. Both hydrophilic and hydrophobic membranes were used. For the hydrophilic membrane, protein deposition depended on electrostatic forces, exhibiting little or no fouling when the proteins and the membrane had charge of equal sign. In the case of the hydrophobic membrane, protein-membrane attractive forces were enough strong to cause β-lactoglobulin deposition even in the presence of repulsive electrostatic forces. Finally, Brisson *el al.* (2007) applied an external electric field during the microfiltration of LF and a whey protein concentrate through an organic membrane. It was observed that the filtration of the mixture deteriorated the transmission of LF and two mayor whey proteins α-lactalbumin and β-lactoglobulin in comparison with the values obtained when these proteins were filtered alone due to the interaction between the positive charged LF and the negative charged β-lactoglobulin at the studied pH. Separation factors obtained between LF and two main whey proteins α-lactalbumin and β-lactoglobulin were 9.1 and 3.0 respectively. An increase in electrical field increased permeate flux by a factor 3, suggesting an important decrease of membrane fouling. Finally, the separation between iron-saturated LF and α-lactalbumin and β-lactoglobuin was done, separation factors in this case were up to 62.4 and 6.7 respectively.

As has been shown, there are many research works which demonstrate the viability of using membrane technology for whey protein fractionation, for this reason, this technique has been used with real multicomponent mixtures: sweet or acid whey or whey protein concentrates. Thus, Lucas *et al.* (1998) investigated the selective extraction of α-lactalbumin from a whey protein concentrate by employing positive charged inorganic membranes. The transmission was studied at pH 7 as a function of the ionic strength. The best results were obtained at low ionic strength with a low β-lactoglobulin transmission and selectivities close to 10. In 1999, Muller *et al.* made a study about the influence of different operation modes on the purity and recovery of α-lactalbumin from acid whey by ultrafiltration. The operation modes employed were: discontinuous concentration, continuous concentration, continuous diafiltration and a combined continuous concentration-diafiltration mode. Results showed that with the combined method, an α-lactalbumin fraction with good purity-yield ratio was obtained in the permeate.

Fractionation methods have also been developed, the majority of them use a two stages strategy: Cordle *et al.* (1990) developed a process which employed metallic oxide membranes. In the first fractionation stage pH 5 was used, achieving the retention of BSA and inmunoglubulins because both α-lactalbumin and β-lactoglobulin, almost discharged, were collected on the permeate. Then, pH was risen up to 7, with this conditions inmunoglubulins go through the membrane. The final permeate has an inmonuglubulins content of 14% in comparison with 7.5% in the initial whey. The following year, Bottomley obtained an α-lactalbumin enriched fraction from sweet whey. In the first stage, the separation of α-lactalbumin was accomplished by using an organic membrane. Second stage was used for the concentration of the α-lactalbumin enriched permeate. Final product presented a α-lactalbumin:β-lactoglobulin ratio of 3:1. Muller *et al.* (2003a y 2003b) proposed a purification method for α-lactalbumin from an acid whey protein concentrate. In their first work, they studied first stage which consisted on prepurification by using ultrafiltration ceramic membranes for total retention of BSA and inmunoglobulins with a limited transmission of β-lactoglobulin. With this stage, the α-lactalbumin purity in permeate was increased from around 0.25 in the initial feed to around 0.44 with a yield close to 0.53. In the second work, they study the second stage, two options were investigated: ultrafiltration and precipitation. With the ultrafiltration stage, the α-lactalbumin enriched permeate was obtained with a purity of 0.65 but the yield was poor 0.15. With the precipitation stage, purity was in the range 0.77-0.99 with a yield between 0.46-0.83. Finally, Cheang and Zydney in 2004 studied a strategy for the purification of both α-lactalbumin and β-lactoglobulin from a whey protein isolate. Separation was achieved using organic membranes and continuous diafiltration mode. Results showed a purification for α-lactalbumin greater than 10-fold with a yield of 95%. The β-lactoglobulin yield was 70% with a purification factor of 8.

In other research works three-stage strategies for whey protein fractionation have been developed. These works can be sorted in two groups according to the final stage which can be HPTFF or ionic exchange chromatography.

In the first group, it can be found the work of Mehra and Kelly (2004), these authors proposed a system for the sequential fractionation of whey proteins based on the modification of pH and ionic strength in order to modify proteins charge and change their hydrodynamic volume. First stage produced a retentate enriched on α-lactalbumin and β-lactoglobulin and a retentate which contained IgG, LF, lactoperoxidase and BSA. In the second stage, permeate of the first stage was fractionated; the new permeate contained α-lactalbumin and the retentate β-lactoglobulin. With the last stage a permeate enriched on IgG or LF and lactoperoxidese was obtained as a function of the selected pH.

In the second group, we can find the following works: Bhattacharjee *et al.* (2006) fractionated an acid whey protein concentrate using two-stage ultrafiltration with organic membranes followed by ion-exchange membrane chromatography. In the first stage, they isolated α-lactalbumin and β-lactoglobulin from BSA, LF and inmunoglobulings which were retained. Then, a concentration process of the α-lactalbumin and β-lactoglobulin fraction was accomplished. Finally, ion-exchange membrane chromatography was used with the retentate of the second stage obtaining a β-lactoglobulin purity of 87.6% on total protein basis. However, this purity was relatively low (26%) on total solids basis. Finally, in 2007, Lu *et al.* studied the obtaining of lactoferrin from bovine calostrum by a two-stage ultrafiltration process with organic membranes and then a purification stage by cation exchange chromatography. In the ultrafiltration stages, transmembrane pressure, tangential velocity and operation temperature were optimized. The product of the ultrafiltration stages was an enriched fraction of LF with a purity of 31% and a yield of 94 %. After the cation exchange chromatography, a final purity and yield of 94% and 82% respectively was achieved.

All the above works illustrate the importance of the development of membrane processes for whey protein fractionation. In this chapter, results obtained in the study of the pH effect on the ultrafiltration of acid whey are presented. This work shows that this process variable is crucial since different responses have been obtained for permeate flow and protein transmission in pH range 3 to 10. These results are very useful for the design of whey ultrafiltration strategies.

Chapter 3

EXPERIMENTAL

PREPARATION OF CLARIFIED ACID WHEY

Acid whey was produced from whole milk. To do this, the milk was centrifuged at 4500 rpm and 4 °C for 30 minutes for fat removal. Then, the pH of the skimmed milk was reduced to 4.2 to produce the coagulation of casein which was also separated by centrifugation under the same conditions as the previous one. To improve permeate flux during the assays, whey was clarified using the method proposed by Rinn *et al.* (1990) and Gesan *et al.* (1995). This method has 7 steps (Figure 9):

- Whey cooling until a temperature between 2-5 °C.
- Addition of 1.2 g/L of $CaCl_2$ to adjust the calcium content.
- pH setting at 7.3 by NaOH.
- Whey heating as fast as possible until 55 °C.
- Whey is maintained at this temperature and pH for 8 min. This allows the aggregation of the lipidic complexes with the particles of calcium phosphate
- Cooling until 10 °C.
- Centrifugation at 4500 rpm for 30 minutes. Thus, the precipitated fraction, which contains the aggregates, and the clarified acid whey (supernatant) are separated.

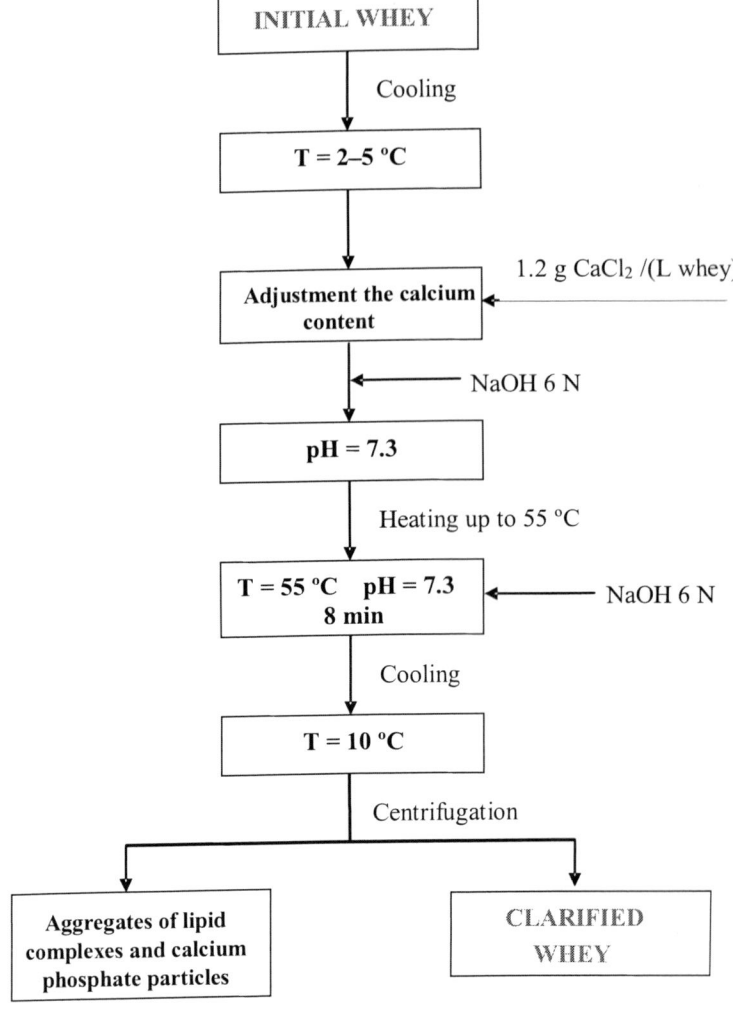

Figure 9. Diagram of the pretreatment applied to acid whey in order to obtain clarified whey.

Figure 10 shows an image of the whey prior to clarification process (non-clarified whey) and other of whey after the process (clarified whey). It can be seen that clarified whey is translucent and doesn't have particles in suspension. By contrast, non-clarified acid whey is less transparent and has some white particles in suspension.

Figure 10. Photograph of both clarified and unclarified whey.

Thanks to clarification process, permeate flux is improved up to 6 times without changing the protein composition of whey (α-lactalbumin, β-lactoglobulin, BSA, IgG y LF) (Almécija, 2007).

EXPERIMENTAL DEVICE

The experimental device used for whey ultrafiltration is schematized in Figure 11. It consists of a 2L feed tank (1), which has a temperature probe (8) and is immersed in a thermostatic bath. The feed is driven, by a positive displacement pump (2), to the membrane module (4). Two streams are obtained from this module: the retentate stream (solid line) and permeate stream (dashed line). The feed stream has a pressure gauge (3) to measure the pressure at the module entrance. In the retentate stream another pressure gauge is placed (5) to measure the pressure at the outlet of the membrane module. A membrane valve (6) to control the pressure and a magnetic flow meter (7) are also placed in the retentate stream. The permeate stream is collected in the accumulated permeate tank (9) and a solution at working pH and temperature is added to the feed tank (11), with a flow equal to permeate flow (10).

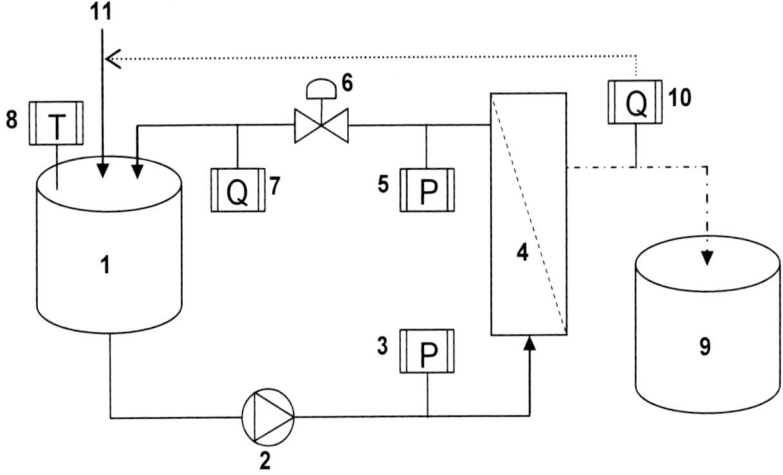

Figure 11. Experimental device used for the continuous diafiltration of clarified acid whey.

The employed membrane was a ceramic membrane model Inside Céram 120 housed in a stainless steel module (Figure 12) and with 300 kDa cut off. This membrane is formed by a tubular support with a diameter of 10 mm and 3 channels with a hydraulic diameter of 3.6 mm. It has a length of 1.2 m and a surface of 0.045 m^2. The support is made of aluminium, titanium and zirconium oxides, whereas titanium oxide is used for the filtering layer. The point of zero charge of Inside Céram membranes is independent of configuration and pore size, depending only on the membrane material (De la Casa, 2006). For the membrane used in this research work, pzc was close to 7.

Figure 12. Module and membranes employed for the continuous diafiltration of clarified acid whey.

The membrane was selected due to its advantages over organic membranes: better resistance to chemical products, it can be used in a wide temperature and pH range, it resists high pressures and its life-time is bigger (Shah *et al.*, 2007). On the other hand, in terms of configuration, tubular module was chosen because it allows an easy scale up (Cui *et al.*, 1997) and, for the specific case of ceramic membranes, this configuration needs less surface than flat configurations in identical conditions (Grangeon and Lescoche, 2000).

EXPERIMENTAL PROCEDURE

The assays were done in continuous diafiltration mode, employing 2 L of clarified acid whey as feeding. Each assay was ended when 4 diavolumes were achieved. pH values between 3 and 10 were tested. The experiment has three different stages:

- Membrane conditioning: Milli-Q water was recirculated at working pH and temperature for 15 min.

- Filtration: Initially, the pH of the feeding was adjusted to the working value, then, a vacuum filtration was done for aggregates removal. Subsequently, the feeding was allocated in the feed tank and the process started. The working temperature was 30 ºC in order to avoid denaturalization problems. Transmembrane pressure was 1 bar for working in the operation zone controlled by the pressure. The circulation velocity was 3.3 m/s which means a retentate flow of 400 L/h, this velocity provides a turbulent flux which minimizes the polarization layer. In continuous diafiltration mode, the permeate flow was collected in an accumulated permeate tank and its time evolution was monitored. With a flow equal to permeate flow, diafiltration water (at working pH and temperature) was added to the feed tank. Diafiltration process ended when 4 diavolume were reached. 1 mL samples of retentate and accumulated permeate were taken each diavolume for protein quantification (α-lactalbumin, β-lactoglobulin, BSA, IgG y LF).
- Cleaning: To achieve membrane regeneration after ultrafiltration, a rinse with distilled water was done in order to eliminate working solution waste. Then, a basic cleaning was done in total recirculation mode at 50 ºC, 350 L/h, 1 bar for 60 minutes for the elimination of organic and biologic fouling. Basic cleaning solution was 20 g/L of NaOH and 2 g/L of sodium docecil sulfate (SDS). Then a rinse with distilled water was done until reach neutrality in both retentate and permeate side. Basic cleaning and the rinse until neutrality were repeated until the complete recovery of the membrane. Only for difficult cases, i.e., if after three basic cleanings the membrane was not recovered, an acid cleaning was accomplished in order to eliminate inorganic fouling. For the acid cleaning a 60% nitric acid solution with a concentration of 2.5 mL/L was employed at 50 ºC, 350 L/h, 1 bar for 15 minutes.

ANALYTICAL METHODS

Individual protein concentrations, including α-lactalbumin, β-lactoglobulin, BSA, lactoferrin and Ig-G, were determined by reversed-phase high-performance liquid chromatography (RP-HPLC) using the method described by Elgar et al. (2000) and extended by Palmano and Elgar (2002). A 1-mL Resource RPC column was operated at room temperature at a flow-rate of 1 mL/min. Solvent A was 0.1% v/v trifluoroacetic acid (TFA) in

Milli-Q water and solvent B was 0.09 % v/v TFA, 90% v/v acetonitrile in Milli-Q water. The column was equilibrated in 80 % solvent A. The gradient used was: 0-1 min, 20 % B; 1-6 min, 20-40 % B; 6-16 min, 40-45 % B; 16-19 min, 45-50 % B; 19-20 min, 50 % B; 20-23 min, 50-70 % B; 23-24 min, 70-100 % B; 24-25 min, 100 % B; 25-27 min, 100-20 % B; 27-30 min, 20 % B. Detection was by absorbance at 214 nm. Sample injection volumes were 50 µL for initial clarified whey and retentate samples and 100 µL for permeate samples.

Chapter 5

RESULTS AND DISCUSSION

The influence of pH on the whey protein fractionation is shown in terms of evolution of permeate flux, protein transmission and purity improvement of the proteins retained.

COMPOSITION OF THE CLARIFIED WHEY

A RP-HPLC chromatogram of the clarified whey is shown in Figure 13. The main whey proteins (α-lactalbumin and β-lactoglobulin, variants A and B) appeared at elution times of 11.4 and 19.6 min respectively. AU_{214nm} was limited to 0.60 in the y-axis for the purpose of presenting significant areas for the other proteins. The smallest peak corresponded to lactoferrin with an elution time of 15.5 min. Regarding the BSA and IgG, their peaks appeared just before (at 16.9 min) and after (at 22.8 min) the β-lactoglobulin peak, respectively. Taking into account the calibration curves, the concentration for each individual whey protein was obtained as follows: α-lactalbumin, 1.00 g/L; β-lactoglobulin, 2.70 g/L; BSA, 0.1 g/L; IgG, 0.40 g/L; lactoferrin, 0.04 g/L.

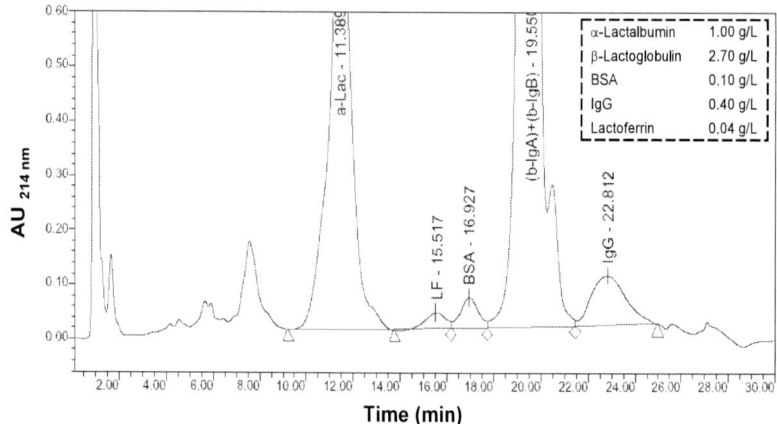

Figure 13. Reverse phase HPLC chromatogram of clarified bovine whey. The concentration of the individual proteins is shown in the text box.

PERMEATE FLUX

The evolution of the permeate flux for the different pH values is shown in Figure 14. Two different behaviours could be observed. First (Figure 14(a)), curves characterized by an initial sharp decrease of flux followed by a flux practically constant until the end of operation. This profile was achieved at pH 4, 5, 6, 7 and 8. The slowest filtrations were obtained at pH=4 and 5 (more than 6 h), in which initial permeate fluxes of 40 and 51 L/(m^{2}h), respectively, decayed to 25 in the first 3 h. In the case of pH 6, 7 and 8, the main decrease (from 63, 65 and 73 L/(m^{2}h) to 43, 42 and 50 L/(m^{2}h), respectively) occurred in the first 2 h with a total time of 3.8, 3.8 and 3.2 h, respectively.

The second kind of curves obtained is shown in Figure 14(b): oscillating or increasing permeate fluxes with time, which were obtained at pH 3, 9 and 10. At the extreme pH values (3 and 10), a linear increase was observed from the initial values of 68 and 89 L/(m^{2}h), reaching 85 and 125 L/(m^{2}h) at the end of the operation, after 2.2 and 1.6 h, respectively. The oscillating curve corresponded to pH 9. In the first hour, flux decayed from 91 to 80 L/(m^{2}h). Finally, a steady flux of 87 L/(m^{2}h) was maintained up to 2.1 h.

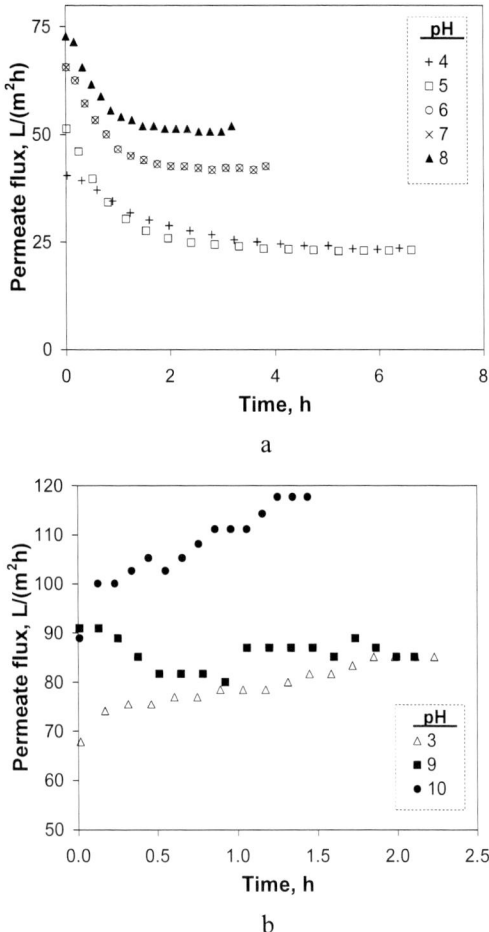

Figure 14. Time evolution of permeate flux at different pH values: (a) permeate flux decreased at the beginning of the operation and then remaining practically constant until the end, and (b) permeate flux increased or oscillated with time.

In order to explain the flux-time profiles in this continuous diafiltration, two opposite phenomena have to be considered. First, total protein concentration decreases during the process provided some protein transmission through the membrane occurs. This promotes an increase in permeate flux. On the other hand, membrane fouling as a consequence of protein adsorption on to the membrane involves a decrease in permeate flux. Both mechanisms (dilution and fouling) depend mainly on the electrostatic interactions protein-protein and protein-membrane, which are relied on the relative position of the work pH respect to the isoelectric points of the proteins (α-lactalbumin, 4.5-4.8; β-lactoglobulin, 5.2; BSA, 4.7-4.9; IgG, 5.5-8.3; lactoferrin, 9.0) (Zydney, 1998) and the point of zero charge of the membrane (7). Therefore, at the extreme pH values (3 and 10), an increase in permeate flux occurs due to the dominance of the dilution in the retentate tank since all the whey proteins and the membrane have the same charge sign (positive and negative, respectively) and, consequently, this repulsion do not favour fouling. On the other hand, at pH 4 and 5 (around the isoelectric points of the most abundant whey proteins) fouling controls owing to the deposition on to the membrane of aggregates of uncharged α-lactalbumin, β-lactoglobulin and BSA molecules. This involves a sharp permeate flux decrease in the first part of the process until a steady flux is reached, probably due to the sweeping effect of the tangential retentate stream.

RETENTATE AND PERMEATE YIELDS OF INDIVIDUAL PROTEINS

The retentate and permeate yields of each whey protein analysed were monitored during the 4 diavolumes of operation for each pH assayed for the purpose of evaluating the sieving characteristics of the membrane. These yields were calculated as the ratio between the mass of protein in the instantaneous retentate and cumulated permeate, respectively, and the mass of protein in the initial feed.

α-LACTALBUMIN AND β-LACTOGLOBULIN

The retentate and permeate yields of α-lactalbumin as a function of the number of diavolumes are shown in Figure 15. In all cases, the biggest variation in yield was obtained in the first 2 diavolumes, with residual changes only in the others. This could be due to membrane fouling, which avoids any protein transport through the membrane at the end of the operation. It can be seen that the curves for retentate and permeate are almost symmetrical with respect to an imaginary horizontal line at a yield value of 0.5. Three different behaviours were observed. For pH=4 and 5, a very low permeate yield was achieved, involving that practically all the initial α-lactalbumin remained in the retentate. On the contrary, the highest permeate yields were achieved at pH=7, 8 and 9 since more than 50% of the original protein was obtained in the cumulated permeate. For pH=3, 6 and 10, intermediate results were observed since more than 40% but less 50% of the original α-lactalbumin passed through the membrane.

The yields of β-lactoglobulin are represented in Figure 16. Again, symmetrical curves were obtained, the main change occurred in the first 2 diavolumes and no significant protein transmission was observed at pH=4 and 5. In this case, the highest transmissions were achieved at pH=3, 7, 8 and 9, in which up to 33% of the β-lactoglobulin present in the initial feed was collected in the cumulated permeate. Lower permeate yields (20%) occurred at pH=6 and 10.

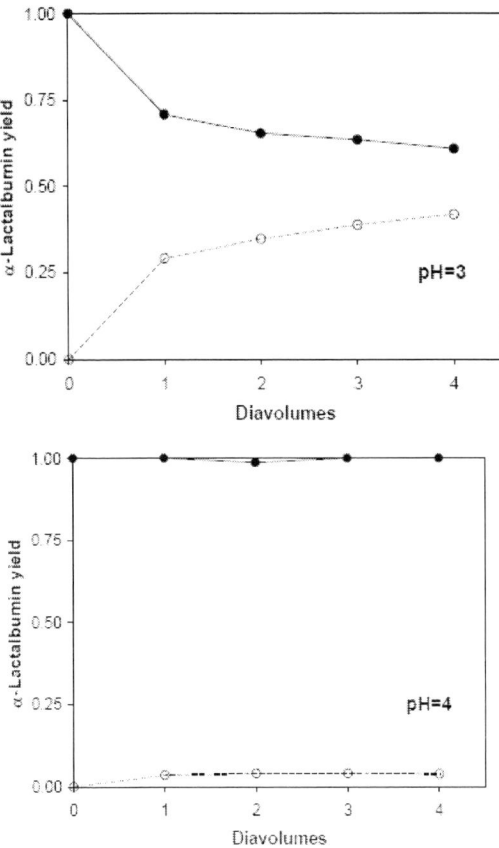

Figure 15. (Continued).

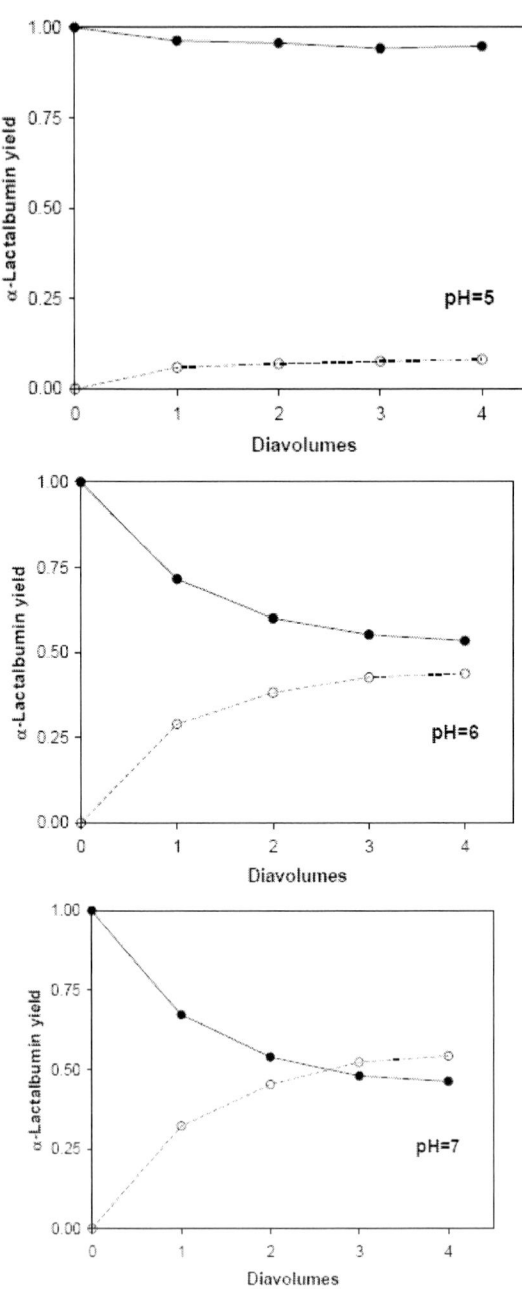

Figure 15. (Continued).

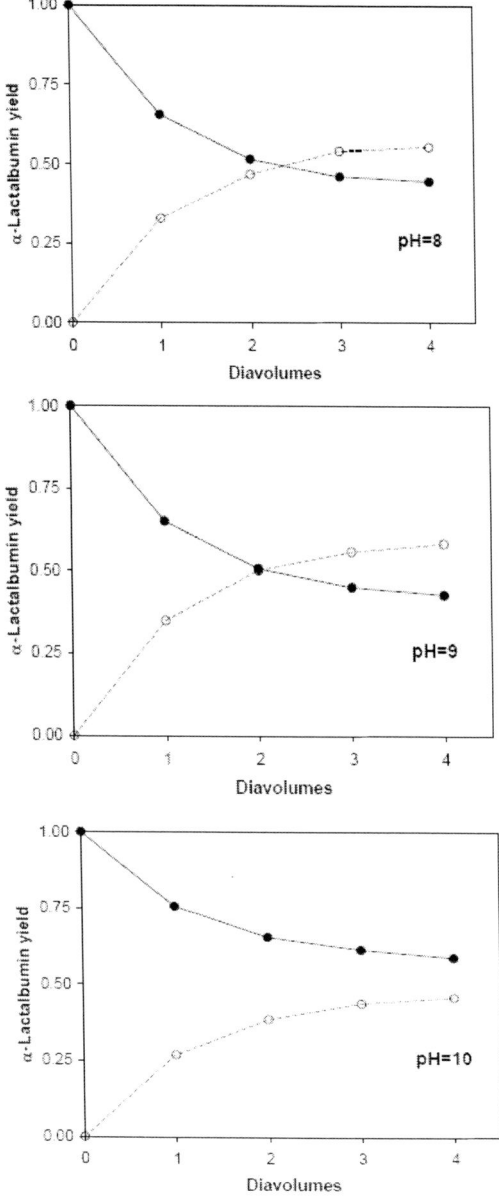

Figure 15. Retentate (●) and permeate (○) yield of α-lactalbumin as a function of the number of diavolumes at different pH values for the continuous diafiltration of clarified bovine whey through a 300 kDa ceramic membrane.

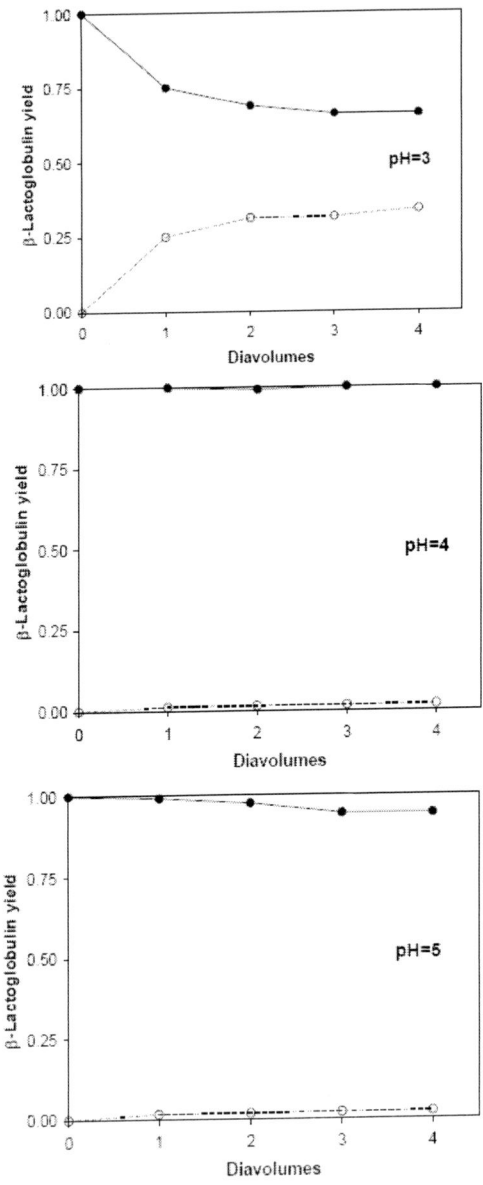

Figure 16. (Continued).

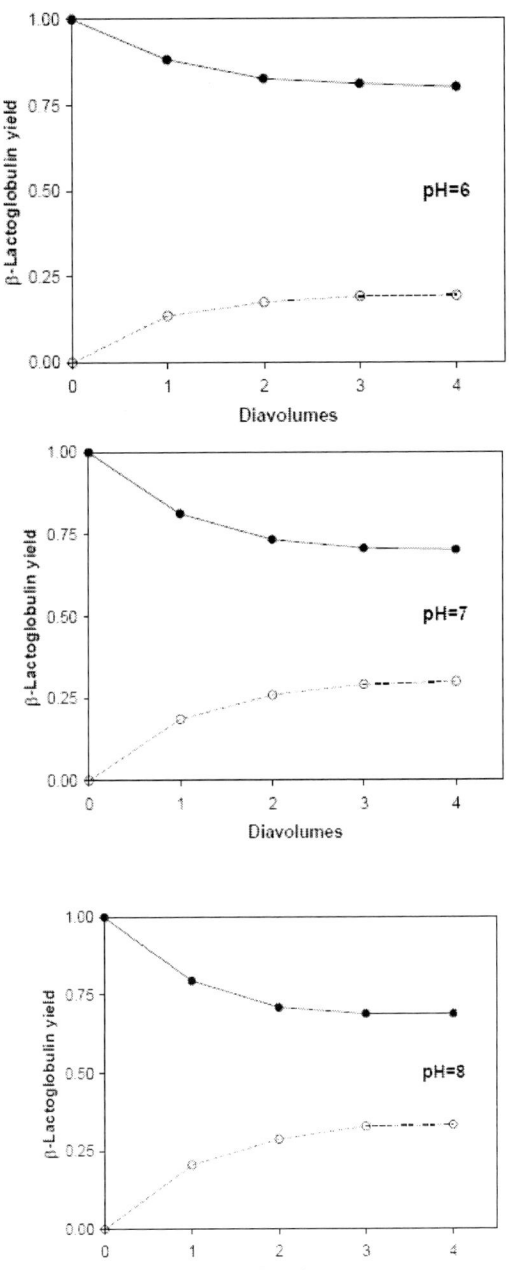

Figure 16. (Continued).

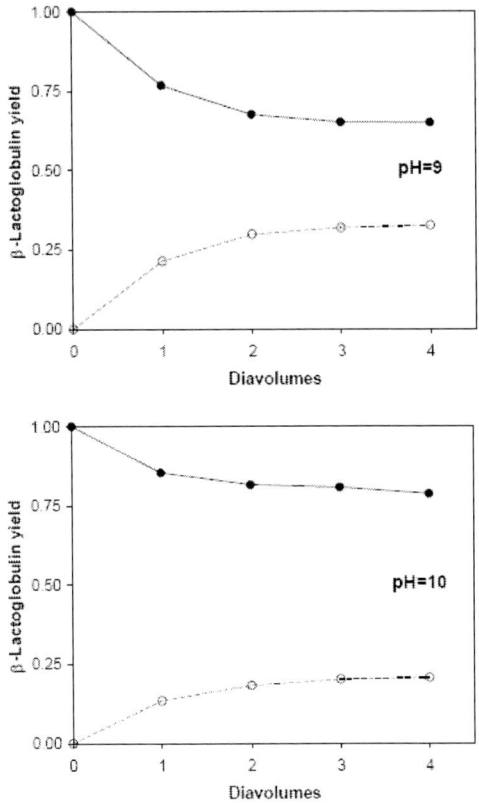

Figure 16. Retentate (●) and permeate (○) yield of β-lactoglobulin as a function of the number of diavolumes at different pH values for the continuous diafiltration of clarified bovine whey through a 300 kDa ceramic membrane.

Taking into account the ratio between the pore size of the membrane (300 kDa) and the molecular weight of the α-lactalbumin and β-lactoglobulin monomer (14 and 18 kDa, respectively), both proteins should pass through the membrane with ease. However, a variety of permeate yields was obtained at changes of the electrostatic environment. This could be due to the conformational changes of the protein molecules are pH-dependent. For instance, at pH=4 and 5, large aggregates of the uncharged proteins of up to 8 molecules are formed (Boye et al., 1997; Taulier and Chalikian, 2001) involving very low transmissions. At pH in the 7-9 interval, around the point of zero charge of the membrane, interactions between the proteins and the membrane are minimised, which results in the highest permeate yields. For the extreme pH values (3 and 10), when protein and membrane repel each other, transmissions were greater than expected, probably due to the importance of convective transport of solute as a consequence of the high permeate flux. A special case occurs for β-lactoglobulin at pH=3, where it is present as a monomer and, therefore, shows permeate yields as high as those obtained in the range 7-9.

BSA, IMMUNOGLOBULINS AND LACTOFERRIN

The yields of BSA are shown in Figure 17. For all the pH assayed, practically all the initial BSA remained in the retentate during the 4 diavolumes.

In Figure 18 it is represented the retentate and permeate yields of IgG. For pH=4, 8, 9 and 10, all the IgG remained in the retentate. For pH=3, 6 and 7, symmetrical curves with high retentate yields (85-89%) were obtained. A special behaviour was observed at pH=5. Although null permeate yield was obtained, only 53 % of the original IgG stayed in the retentate at the end of the diafiltration.

The yields of lactoferrin are presented in Figure 19. Null permeate yields were obtained in all cases. At the extreme pH values (3 and 10) the retentate yield was 100 %. In the experiment at its isoelectric point, pH=9, the final retentate yield was high (91%). However, in the rest of the experiments, retentate yields were significantly lower, mainly at pH=5 and 7, in which only 26% of the original lactoferrin remained in the final retentate.

Therefore, these proteins were much more retained than α-lactalbumin and β-lactoglobulin due to their higher molecular size (BSA, 69 kDa; IgG, 150-1000 kDa; LF, 78 kDa).

Regarding the loss of significant amounts of IgG at pH=5 and lactoferrin at pH=4-9 could be due to 3 possible causes: (1) protein adsorption to the membrane; (2) protein denaturation by shear stress caused by the circulation of the retentate stream at high velocities (Cheryan, 1998) and (3) association with one of the major whey proteins. For example, lactoferrin forms noncovalent complexes with β-lactoglobulin or BSA with molar ratios 2:1 and 1:1, respectively (Lampreave *et al.*, 1990), provided no protein-protein repulsion occurs. The diafiltration process could favour this association owing to the elution of salt ions in solution, which are in equilibrium with ions joined to local charges of the proteins.

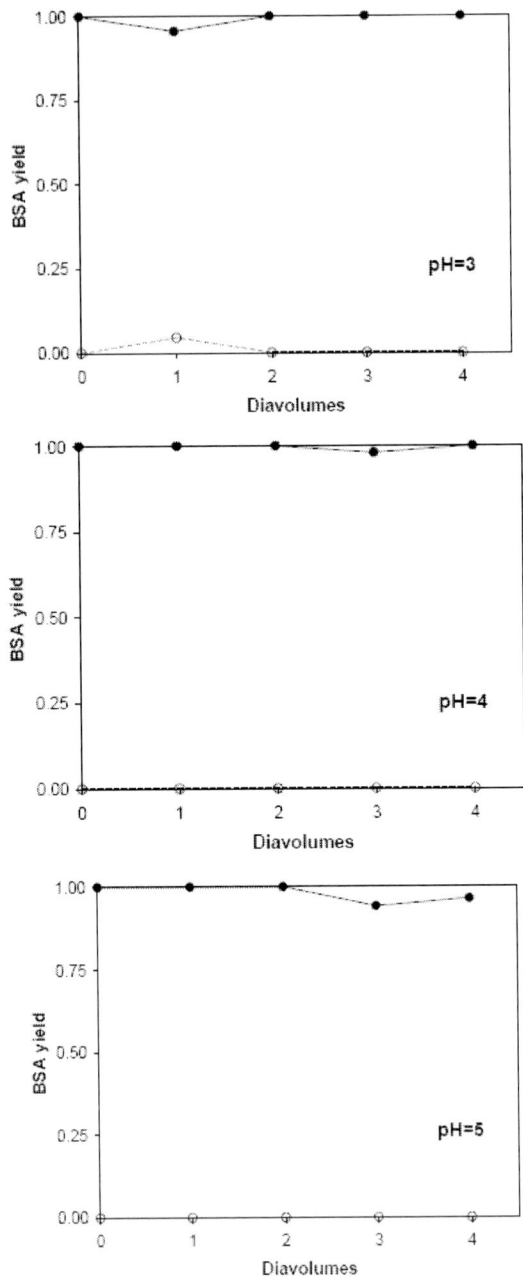

Figure 17. (Continued).

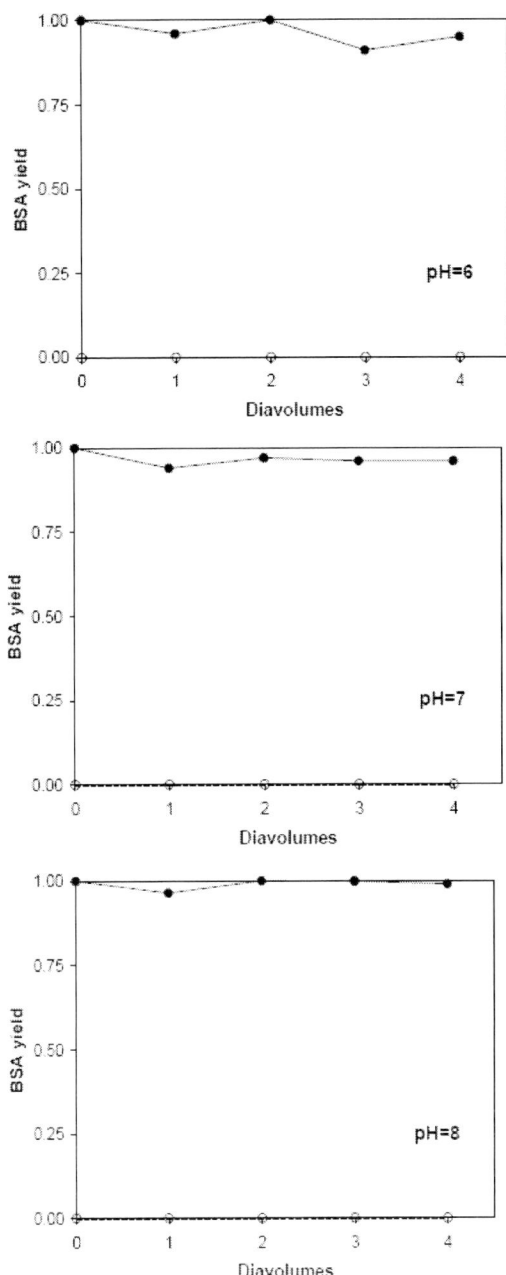

Figure 17. (Continued).

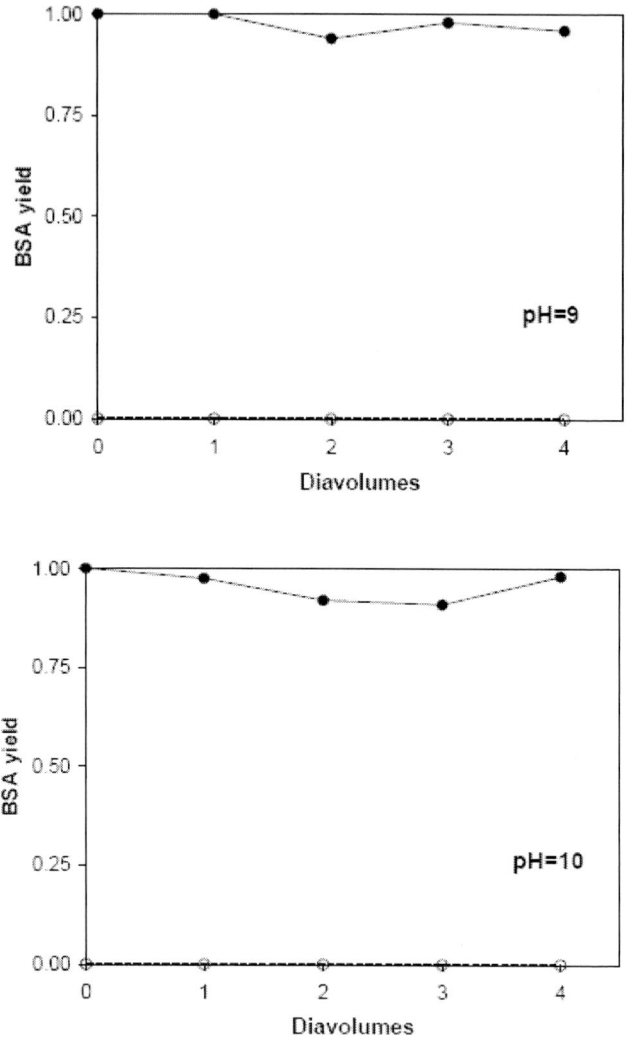

Figure 17. Retentate (●) and permeate (○) yield of bovine serum albumin as a function of the number of diavolumes at different pH values for the continuous diafiltration of clarified bovine whey through a 300 kDa ceramic membrane.

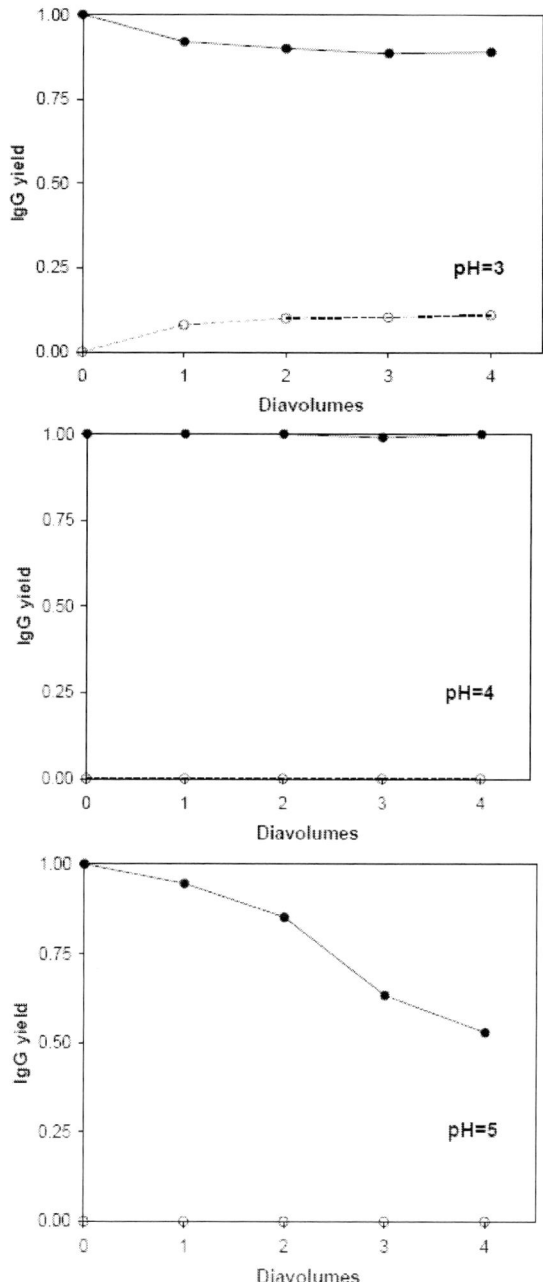

Figure 18. (Continued).

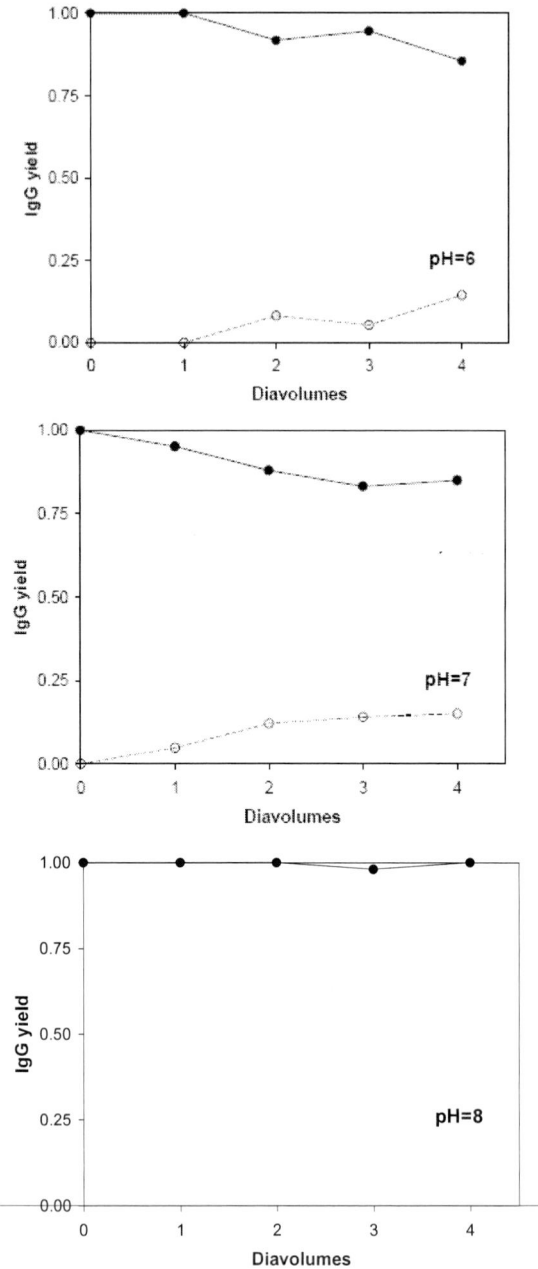

Figure 18. (Continued).

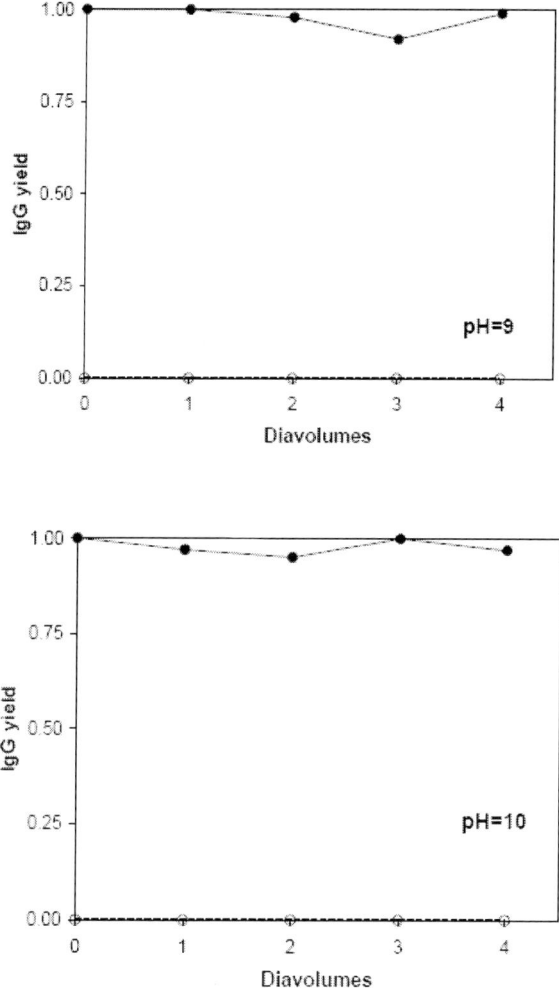

Figure 18. Retentate (●) and permeate (○) yield of immunoglobulin G as a function of the number of diavolumes at different pH values for the continuous diafiltration of clarified bovine whey through a 300 kDa ceramic membrane.

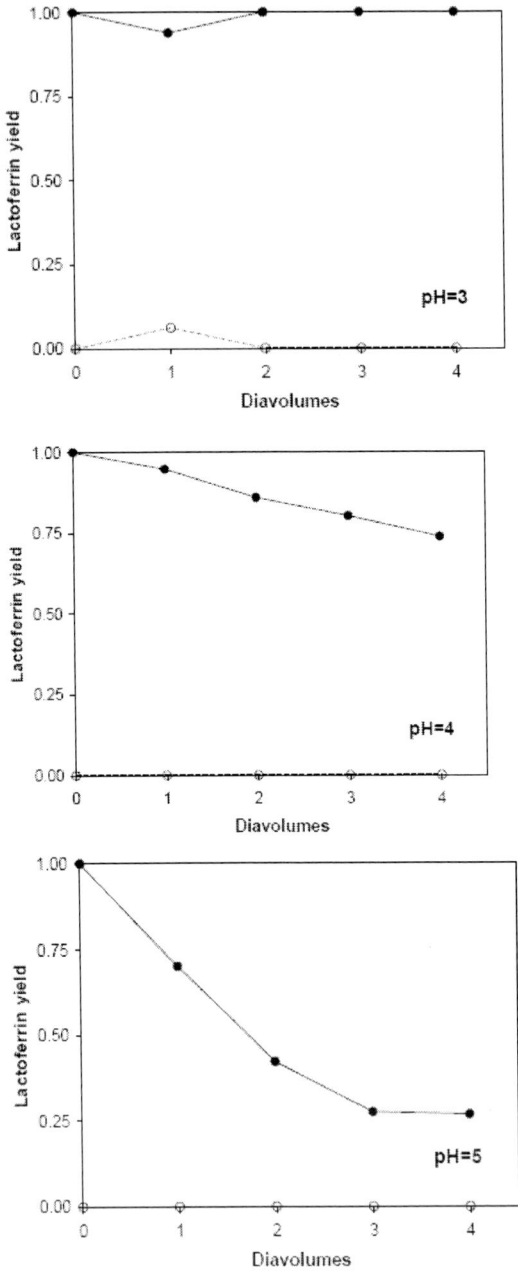

Figure 19. (Continued).

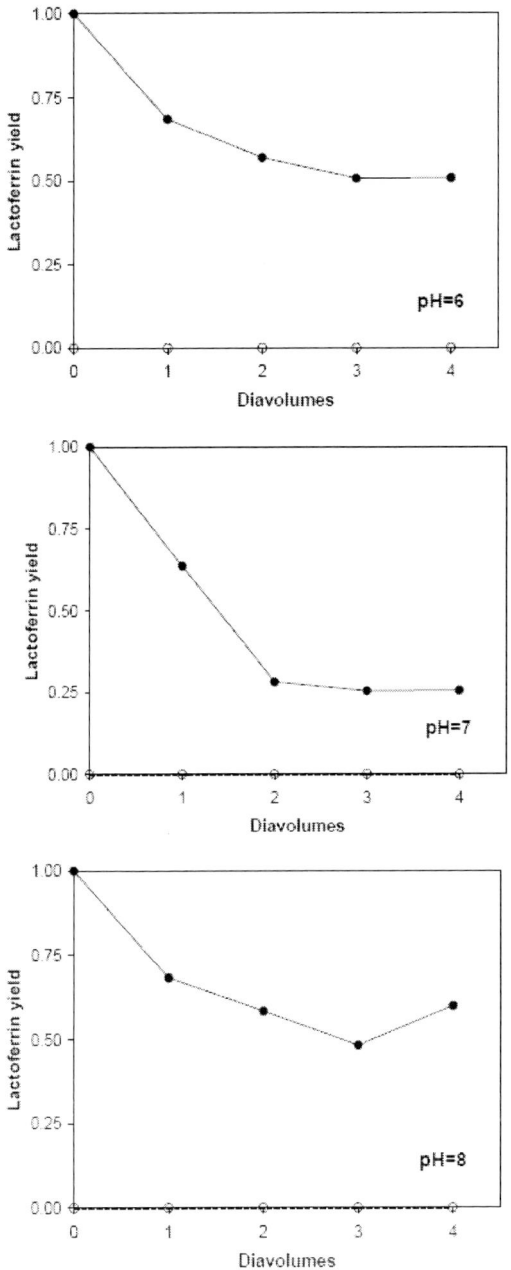

Figure 19. (Continued).

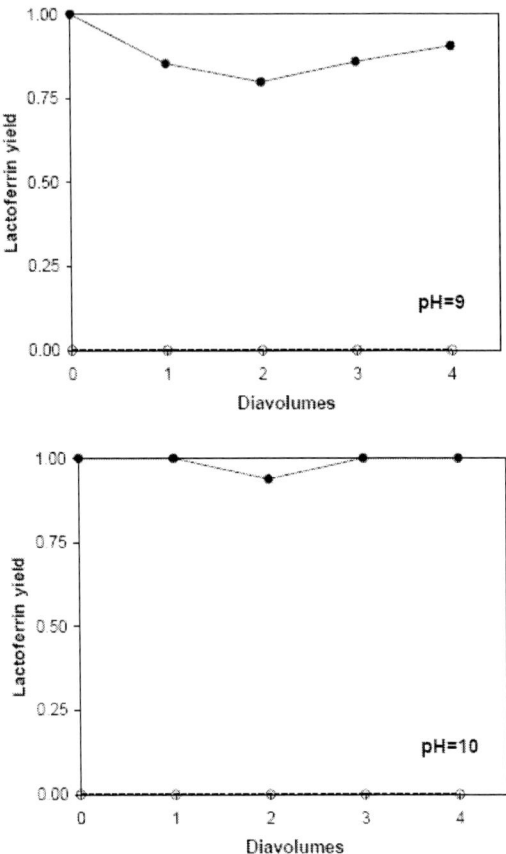

Figure 19. Retentate (●) and permeate (○) yield of lactoferrin as a function of the number of diavolumes at different pH values for the continuous diafiltration of clarified bovine whey through a 300 kDa ceramic membrane.

PURITY IMPROVEMENT OF THE PROTEINS RETAINED

The results obtained indicate that, except for pH=4 and 5, important amounts of α-lactalbumin and β-lactoglobulin passed through the membrane, whereas the rest of the proteins studied were preferently retained. This suggests that the 300 kDa membrane could be used to separate the whey proteins in two fractions: (1) α-lactalbumin and β-lactoglobulin in the permeate and (2) BSA, IgG and lactoferrin in the retentate. In order to evaluate qualitatively this possibility, the improvement of purity experienced by the proteins retained was calculated.

The purity in the original whey for BSA, IgG and lactoferrin was calculated as the ratio between the concentration of individual protein and the sum of the concentration of the five proteins. The values obtained were: BSA, 2.4%; IgG, 9.4%; and LF, 0.9%. In the same way, the final purity, i.e., at the end of the diafiltration process was calculated considering the concentrations at diavolume=4. Finally, the purity improvement factor was the ratio between final and initial purity. Therefore, factor values above 1 involve to some extent a satisfactory process performance since the protein of interest is purer than in the original whey. On the other hand, values below 1 indicate a negative effect of the ultrafiltration since the protein purity decreases respect to the initial one.

Purity improvement factors for BSA, IgG and lactoferrin for each pH assayed are shown in Table 2. For BSA, some purity improvement was achieved in all cases, with maximum values of 1.5 at pH=3, 7, 8 and 9. In the case of IgG, the factor decreased from 1.3 at pH= 3 to 0.6 at pH=5 and then increased up to 1.6 at an optimum pH of 9. Finally, some improvements in purity of lactoferrin were obtained only at the extreme pH values, with two maxima increments of 50% at pH=3 and 9. Therefore, if our aim were to fractionate the original whey in the two parts indicated above with the 300 kDa membrane, working pH should be adjusted to 3 or 9. Moreover, both values are equally appropriate since almost identical filtration times were needed to complete the 4 diavolumes (2.2 and 2.1 h, respectively).

Table 2. Purity improvement factor achieved for the retained proteins (bovine serum albumin, immunoglobulin G and lactoferrin) as a function of pH in the continuous diafiltration up to 4 diavolumes of clarified bovine whey through a 300 kDa ceramic membrane

Retained protein	pH							
	3	4	5	6	7	8	9	10
BSA	1.5	1.0	1.1	1.3	1.5	1.5	1.5	1.3
IgG	1.3	1.0	0.6	1.1	1.3	1.5	1.6	1.3
LF	1.5	0.7	0.3	0.7	0.4	0.9	1.5	1.3

CONCLUSION

Some conclusions could be obtained from the research work shown, where the effect of pH on the fractionation of whey proteins has been studied:

- The lower permeate fluxes were obtained at pH=4 and 5, i.e., around the isoelectric points of α-lactalbumin, β-lactoglobulin and BSA. It could be due to the membrane fouling, probably caused by the deposition of aggregates of the uncharged protein molecules, which involves a sharp decrease of flux in the first two diavolumes of operation.
- At the extreme pH values, 3 and 10, protein-protein and protein-membrane repulsions make fouling difficult and, therefore, the dilution mechanism dominates, involving permeate fluxes which increase linearly with time.
- The smallest proteins, α-lactalbumin and β-lactoglobulin, were eluted through the membrane. Maximum permeate yields were achieved around the point of zero charge of the membrane, 7, with values up to 58% and 33% for α-lactalbumin and β-lactoglobulin, respectively.
- The proteins with higher molecular weight, BSA, IgG and lactoferrin, were preferently retained. In the case of BSA and lactoferrin, their original purity could be multiplied by a factor 1.5 when operating at pH=3 and 9. Similarly, the purity of IgG was improved 1.6 times at pH=9.

ACKNOWLEDGMENTS

This research was supported by the Spanish Plan Nacional I+D+I, under the Projects PPQ-2002-02235 and CTQ-2005-02653.

REFERENCES

Almécija, M.C. (2007). Obtención de lactoferrina bovina mediante ultrafiltración de lactosuero. *Doctoral Thesis.* (in Spanish)

Álvarez, C., Bertorello, H., Strumla, M. and Sánchez E. I. (1996). Preparation and characterization of new biospecific adsorbents with fatty acids as ligands, usable to retain bovine serum albumin, *Polymer*, 37, 3715.

Baker, E.N. and Baker, H.M. (2005). Molecular structure, binding properties and dynamics of lactoferrin, *CMLS-Cell. Mol. Life Sci.*, 62, 2531.

Beulens, J.W.J., Bindels, J.G., de Graaf, C., Alles, M.S. and Wouters-Wesseling, W. (2004). Alpha-lactalbumin combined with a regular diet increases plasma Trp-LNAA ratio, *Physiol. Behav.*, 81, 585.

Bhattacharjee, S., Bhattacharjee, C. and Datta, S. (2006). Studies on the fractionation of β-lactoglobulin from casein whey using ultrafiltration and ion-exchange membrane chromatography, *J. Membrane Sci.*, 275, 141.

Bottomley, R.C. (1991). Process for obtaining concentrates having a high alpha-lactalbumin content from whey, *US Patent.* 5,008,376.

Boye, J.I., Alli, I. and Ismail, A.A. (1997). Use of differential scanning calorimetry and infrared spectroscopy in the study of thermal and structural stability of α-lactalbumin, *J. Agr. Food Chem.*, 45, 1116.

Brisson, G., Britten, M. and Pouliot, Y. (2007). Electrically-enhanced crossflow microfiltration for separation of lactoferrin from whey protein mixtures, *J. Membrane Sci.*, 297, 206.

Brock, J.H. (2002). The physiology of lactoferrin, *Biochem. Cell Biol.*, 80, 1.

Casal, E., Montilla, A., Moreno, F.J., Olano, A. and Corzo, N. (2006). Use of chitosan for selective removal of β-lactoglobulin from whey, *J. Dairy Sci.*, 89, 1384.

Chan, R., Chen, V. and Bucknall, M.P. (2004). Quantitative analysis of membrane fouling by protein mixtures using MALDI-MS, *Biotechnol. Bioeng.*, 85, 190.

Chatterton, D.E.W., Smithers, G., Roupas, P. and Brodkorb, A. (2006). Bioactivity of β-lactoglobulin and α-lactalbumin- Technological implications for processing, *Int. Dairy J.*, 16, 1229.

Cheang, B. and Zydney, A.L. (2003). Separation of α-lactalbumin and β-lactoglobulin using membrane ultrafiltration, *Biotechnol. Bioeng.*, 83, 201.

Cheang, B. and Zydney, A.L. (2004). A two-stage ultrafiltration process for fractionation of whey protein isolate, *J. Membrane Sci.*, 231, 159.

Cherkasov, A.N. and Polotsky, A.E. (1996). The resolving power of ultrafiltration. *J. Membrane Sci.* 110, 79-82.

Cheryan, M. (1998). Ultrafiltration and microfiltration handbook, Technomic Publishing Company, Lancaster.

Christiansen, K.F., Vegarud, G., Langsrud, T., Ellekjaer, M.R. and Egelandsdal, B. (2004). Hydrolyzed whey proteins as emulsifiers and stabilizers in high-pressure processed dressings, *Food Hydrocolloids*, 18, 757.

Conrado, L.S., Veredas, V., Nóbrega, E.S. and Santana, C.C. (2005). Concentration of α-lactalbumin from cow milk whey through expanded bed adsorption using a hydrophobic resin, *Braz. J. Chem. Eng.*, 22, 501.

Cordle, C.T., Thomas, R.L., Criswell, L.G. (1990). Enrichment and concentration of proteins by ultrafiltration, *US Patent.* 4,897,465.

Cui, Z.F., Bellara, S.R. and Homewood, P. (1997). Airlift crossflow membrane filtration - A feasibility study with dextran ultrafiltration, *J. Membrane Sci.*, 128, 83.

De la Casa, E.J. (2006). Estudio de las interacciones proteína-membrana en los procesos de filtración tangencial. *Doctoral Thesis.* (in Spanish)

De la Casa, E.J., Guadix, A., Ibáñez, R. and Guadix, E.M. (2007). Influence of pH and salt concentration on the cross-flow microfiltration of BSA through a ceramic membrane, *Biochem. Eng. J.*, 33, 110.

Dunlap, C.A. and Côté, G.L. (2005). β-lactoglobulin-Dextran Conjugates: Effect of Polysaccharide Size on Emulsion Stability, *J. Agr. Food Chem.*, 53, 419.

Ekici, P., Backleh-Sohrt, M. and Parlar, H. (2005). High efficiency enrichment of total and single whey proteins by pH controlled foam fractionation, *Int. J. Food Sci. Nutr.*, 56, 223.

Elgar, D.F., Norris, C.S., Ayers, J.S., Pritchard, M., Otter, D.E. and Palmano, K.P. (2000). Simultaneous separation and quantitation of the major bovine whey proteins including proteose peptone and caseinomacropeptide by reversed-phase high-performance liquid chromatography on polystyrene-divinylbenzene, *J. Chromatogr. A,* 878, 183.

Firebaugh, J.D. and Daubert, C.R. (2005) Emulsifying and foaming properties of a derivatized whey protein ingredient, *Int. J. Food Prop.,* 8, 243.

FitzGerald, R.J. and Meisel, H. (1999). Lactokinins: whey protein-derived ACE inhibitory peptides, *Nahrung,* 43, 165.

Foster, J.F. (1977). Albumin structure, function and uses (V. M. Rosenoer, M. Oratz, and M. A. Rothschild, eds), Pergamon, Oxford, 53.

Fuda, E., Jauregi, P. and Pyle, D.L. (2004). Recovery of lactoferrin and lactoperoxidase from sweet whey using colloidal gas aphrons (CGAs) generated from an anionic surfactant, *AOT, Biotechnol. Progr.,* 20, 514.

Fuda, E., Bhatia, D., Pyle, D.L. and Jauregi, P. (2005). Selective separation of β-lactoglobulin from sweet whey using CGAs generated from the cationic surfactant CTAB, *Biotechnol. Bioeng.,* 90, 532.

Gesan, G., Daufin, G., Merin, U., Labbe, J.P. and Quemerais, A. (1995). Microfiltration performance: physicochemical aspects of whey pretreatment, *J. Dairy Res.,* 62, 269.

Ghosh, R. and Cui, Z.F. (2000a). Protein purification by ultrafiltration with pre-treated membrane, *J. Membrane Sci.,* 167, 47.

Ghosh, R. and Cui, Z.F. (2000b). Simulation study of the fractionation of proteins using ultrafiltration, *J. Membrane Sci.,* 180, 29.

Glaser, L.A., Paulson, A.T., Speers, R.A., Yada, R.Y. and Rousseau, D. (2007). Foaming behaviour of mixed bovine serum albumin–protamine systems, *Food Hydrocolloid.,* 21, 495.

Grangeon, A., Lescoche, P. (2000). Flat ceramic membranes for the treatment of dairy products: comparison with tubular ceramic membranes, *Lait,* 80, 5.

Grybos, J., Marszalek, M., Lekka, M., Heinrich, F. and Troger, W. (2004). PAC studies of BSA conformational changes, *Hyperfine Interact.,* 159, 323.

Herceg, Z., Rezek, A., Lelas, V., Kresic, G. and Franetovic, M. (2007). Effect of carbohydrates on the emulsifying, foaming and freezing properties of whey protein suspensions, *J. Food Eng.,* 79, 279.

Howell, J.A., Wu, D. and Field, R.W. (1999). Transmission of bovine albumin under controlled flux ultrafiltration, *J. Membrane Sci.,* 152, 117.

Ibáñez, R., Almécija, M.C., Guadix, A. and Guadix, E.M. (2007). Dynamics of the ceramic ultrafiltration of model proteins with different isoelectric point: Comparison of β-lactoglobulin and lysozyme, *Sep. Purif. Technol.*, 57, 314.
Kouoh, F., Gressier, B., Luyckx, M., Brunet, C., Dine, T., Cazin, M. and Cazin, J.C. (1999). Antioxidant properties of albumin: Effect on oxidative metabolism of human neutrophil granulocytes, *Il Farmaco*, 54, 695.
Kushibiki, S., Hodate, K., Kurisaki, J., Shingu, H., Ueda, Y., Watanabe, A. and Shinoda, M. (2001). Effect of β-lactoglobulin on plasma retinol and triglyceride concentrations, and fatty acid composition in calves, *J. Dairy Res.*, 68, 579.
Lampreave, F., Pineiro, A., Brock, J.H., Castillo, H., Sanchez, L. and Calvo, M. (1990). Interaction of bovine lactoferrin with other proteins of milk whey, *Int. J. Biol. Macromol.*, 12, 2.
Legrand, D., Elass, E., Carpentier, M. and Mazurier, J. (2005). Lactoferrin: a modulator of immune and inflammatory responses, *Cell. Mol. Life Sci.*, 62, 2549.
Legrand, D., Elass, E., Carpentier, M. and Mazurier, J. (2006). Interactions of lactoferrin with cells involved in immune function, *Biochem. Cell Biol.*, 84, 282.
Lien, E.L. (2003). Infant formulas with increased concentrations of α-lactalbumin, *Am. J. Clin. Nutr.*, 77, 1555.
Lu, R.R., Xu, S.Y., Wang, Z. and Yang, R.J. (2007). Isolation of lactoferrin from bovine colostrum by ultrafiltration coupled with strong cation exchange chromatography on a production scale, *J. Membrane Sci.*, 297, 152.
Lucas, D., Rabiller-Baudry, M., Millesime, L., Chaufer, B. and Daufin, G. (1998). Extraction of α-lactalbumin from whey protein concentrate with modified inorganic membranes, *J. Membrane Sci.*, 148, 1.
Lucena, M.E., Alvarez, S., Menéndez, C., Riera, F.A. and Alvarez, R. (2006). Beta-lactoglobulin removal from whey protein concentrates. Production of milk derivatives as a base for infant formulas, *Sep. Purif. Technol.*, 52, 310.
Markus, C.R., Olivier, B., Panhuysen, G.E.M., Van der Gugten, J., Alles, M.S., Tuiten, A., Westenberg, H.G.M., Fekkes, D., Koppeschaar, H.F. and de Haan, E.E.H.F. (2000). The bovine protein α-lactalbumin increases the plasma ratio of tryptophan to the other large neutral amino acids, and in vulnerable subjects raises brain serotonin activity, reduces

cortisol concentration, and improves mood under stress, *Am. J. Clin. Nutr.*, 71, 1536.
Marshall, A.D., Munro, P.A. and Trägårdh, G. (2003). Influence of ionic calcium concentration on fouling during the cross-flow microfiltration of β-lactoglobulin solutions, *J. Membrane Sci.*, 217, 131.
Mathews, C.K., van Holde, K.E. and Ahern, K.G. (2000) Biochemistry 3 rd. edition, Addison Wesley Longman, Inc. Benjamin/Cummings.
McIntosh, G.H., Royle, P.J., Le Leu, R.K., Regester, G.O., Johnson, M.A., Grinsted, R.L., Kenward, R.S. and Smithers, G.W. (1998). Whey proteins as functional food ingredients?, *Int. Dairy J.*, 8, 425.
Mehra, R. and Kelly, P.M. (2004). A membrane filtration approach to whey protein fractionation, The Irish Scientist Year Book, Oldbury Publishing Limited.
Mehra, R., Marnila, P. and Korhonen, H. (2006). Milk immunoglobulins for health promotion, *Int. Dairy J.*, 16, 1262.
Militello, V., Vetri, V. and Leone, M. (2003). Conformational changes involved in thermal aggregation processes of bovine serum albumin, *Biophys. Chem.*, 105, 133.
Militello, V., Casarino, C., Emanuele, A., Giostra, A., Pullara, F. and Leone, M. (2004). Aggregation kinetics of bovine serum albumin studied by FTIR spectroscopy and light scattering, *Biophys. Chem.*, 107, 175.
Moore, S.A., Anderson, B.F., Groom, C.R., Haridas, M. and Baker, E.N. (1997). Three-dimensional structure of diferric bovine lactoferrin at 2.8 Å resolution, *J. Mol. Biol.*, 274, 222.
Mullally, M.M., Meisel, H. and FitzGerald, R.J. (1997). Angiotensin-I-converting enzyme inhibitory activities of gastric and pancreatic proteinase digests of whey proteins, *Int. Dairy J.*, 7, 299.
Muller, A., Daufin, G. and Chaufer, B. (1999). Ultrafiltration modes of operation for the separation of α-lactalbumin from acid casein whey, *J. Membrane Sci.*, 153, 9.
Muller, A., Chaufer, B., Merin, U. and Daufin, G. (2003a). Prepurification of α-lactalbumin with ultrafiltration ceramic membranes from acid casein whey: *Study of operating conditions, Lait*, 83, 111.
Muller, A., Chaufer, B., Merin, U. and Daufin, G. (2003b). Purification of α-lactalbumin from a prepurified acid whey: *Ultrafiltration or precipitation, Lait*, 83, 439.
Neirynck, N., Van der Meeren, P., Gorbe, S.B., Dierckx, S. and Dewettinck, K. (2004). Improved emulsion stabilizing properties of whey protein isolate by conjugation with pectins, *Food Hydrocolloid.*, 18, 949.

Neyestani, T.R., Djalali, M. and Pezeshki, M. (2003). Isolation of α-lactalbumin, β-lactoglobulin, and bovine serum albumin from cow's milk using gel filtration and anion-exchange chromatography including evaluation of their antigenicity, *Protein Expres. Purif.*, 29, 202.

Nyström, M., Aimar, P., Luque, S., Kulovaara, M. and Metsämuuronen, S. (1998). Fractionation of model proteins using their physiochemical properties, *Colloid Surface A*, 138, 185.

Palmano, K.P. and Elgar, D.F. (2002). Detection and quantitation of lactoferrin in bovine whey samples by reversed-phase high-performance liquid chromatography on polystyrene-divinylbenzene, *J. Chromatogr. A*, 947, 307.

Pérez, M.D., Sánchez, L., Aranda, P., Ena, J.M., Oria, R. and Calvo, M. (1992). Effect of β-lactoglobulin on the activity of pregastric lipase. A possible role for this protein in ruminant milk, *Biochim. Biophys. Acta*, 1123, 151.

Pessela, B.C.C., Torres, R., Batalla, P., Fuentes, M., Mateo, C., Fernández-Lafuente, R. and Guisán, J.M. (2006). Simple purification of immunoglobulins from whey proteins concentrate, *Biotechnol. Progr.*, 22, 590.

Pihlanto-Leppälä, A. (2001). Bioactive peptides derived from bovine whey proteins: opioid and ace-inhibitory peptides, *Trends Food Sci. Technol.*, 11, 347.

Poole, S., West, S.I. and Walters, C.L. (1984). Protein-protein interactions: Their importance in the foaming of heterogeneous protein systems, *J. Sci. Food Agric.*, 35, 701.

Prádanos, P., Hernández, A., Calvo, J.I. and Tejerina, F. (1996). Mechanisms of protein fouling in cross-flow UF through an asymmetric inorganic membrane, *J. Membrane Sci.*, 114, 115.

Rao, S. and Zydney, A.L. (2005). Controlling protein transport in ultrafiltration using small charged ligands, *Biotechnol. Bioeng.*, 91, 733.

Rinn, J.C., Morr, C.V., Seo, A. and Surak, J.G. (1990). Evaluation of nine semi-pilot scale whey pretreatment modifications for producing whey protein concentrate, *J. Food Sci.*, 55, 510.

Saksena, S. and Zydney, A.L. (1994). Effect of solution pH and ionic strength on the separation of albumin from immunoglobulins (IgG) by selective filtration, *Biotechnol. Bioeng.*, 43, 960.

Schlatterer, B., Baeker, R. and Schlatterer, K. (2004). Improved purification of β-lactoglobulin from acid whey by means of ceramic hydroxyapatite chromatography with sodium fluoride as a displacer, *J. Chromatogr. B*, 807, 223.

References

Shah, T.N., Foley, H.C. and Zydney, A.L. (2007). Development and characterization of nanoporous carbon membranes for protein ultrafiltration, *J. Membrane Sci.*, 295, 40.

Stamler, J.S., Singel, D.J. and Loscalzo, J. (1992). Biochemistry of Nitric Oxide and Its Redox-Activated Forms, Science, 258, 1898.

Taulier, N. and Chalikian, T.V. (2001). Characterization of pH-induced transitions of β-lactoglobulin: ultrasonic, densimetric, and spectroscopic studies, *J. Mol. Biol.*, 314, 873.

Tsuda, H., Sekine, K., Fujita, K. and Iigo, M. (2002). Cancer prevention by bovine lactoferrin and underlying mechanisms-a review of experimental and clinical studies, *Biochem. Cell Biol.* 80, 131.

Valenti, P. and Antonini, G. (2005). Lactoferrin: an important host defence against microbial and viral attack, *Cell. Mol. Life Sci.*, 62, 2576.

Van der Kraan, M.I.A., Groenink, J., Nazmi, K., Veerman, E.C.I., Bolscher, J.G.M. and Amerongen, A.V.N. (2004). Lactoferrampin: a novel antimicrobial peptide in the N1-domain of bovine lactoferrin, *Peptides*, 25, 177.

Van Reis, R., Gadam, S., Frautschy, L.N., Orlando, S., Goodrich, E.M., Saksena, S., Kuriyel, R., Simpson, C.M., Pearl, S. and Zydney, A.L. (1997). High performance tangential flow filtration, *Biotechnol. Bioeng.*, 56, 71.

Van Reis, R. and Zydney, A. (2001). Membrane separations in biotechnology, *Curr. Opin. Biotech.*, 12, 208.

Wolman, F.J., Maglio, D.G., Grasselli, M. and Cascone, O. (2007). One-step lactoferrin purification from bovine whey and colostrum by affinity membrane chromatography, *J. Membrane Sci.*, 288, 132.

Yang Jr., F., Zhang, M., Chen, J. and Liang, Y. (2006). Structural changes of α-lactalbumin induced by low pH and oleic acid, *BBA-Proteins Proteom.*, 1764, 1389.

Zulkali, M.M.D., Ahmad, A.L. and Derek, C.J.C. (2005). Membrane application in proteomic studies: Preliminary studies on the effect of pH, ionic strength and pressure on protein fractionation, *Desalination*, 179, 381.

Zydney, A.L. (1998). Protein separations using membrane filtration: new opportunities for whey fractionation, *Int. Dairy J.*, 8, 243.

INDEX

A

absorption, vii, 7, 12, 13
acetonitrile, 33
acid, 5, 6, 9, 11, 18, 22, 24, 25, 26, 27, 28, 30, 31, 32, 64, 65, 66, 67
activated carbon, 2
adsorption, 18, 19, 37, 45, 62
advantages, 31
aggregation, 5, 10, 21, 27, 65
aggregation process, 65
alanine, 7
albumin, 63, 64, 66
aldosterone, 4
alters, 21
aluminium, 30
amines, 15
amino acids, 6, 7, 11, 64
angiotensin I-converting enzyme (ACE), 4
angiotensin II, 4
antigen, 11
antigenicity, 66
antioxidant, 11
antitumor, 4
antiviral properties, vii
apoptosis, 6
appetite, 5
aqueous by-product, vii
arginine, 13
aspartic acid, 7
attachment, 12

B

bacteria, 12, 14
bacterial infection, 14
bacteriostatic, 14
beneficial effect, 7
beverages, 8
bicarbonate, 13
bile, 15
Biological Oxygen Demand (BOD), 1
biotechnology, 20, 67
blood plasma, 11
blood pressure, 4
bonds, 5, 8, 9, 11
bovine serum albumin (BSA), viii
bradykinin, 4
brain, 6, 64
breast milk, 6

C

calcium, 5, 22, 27, 65
calibration, 35
cancer, 15
candidates, 6
carbohydrate, 6

carbohydrates, 63
carbon, 67
carcinogenesis, 15
casein, 1, 6, 27, 61, 65
catalyst, 14
cation, 26, 64
cell membranes, 11
central nervous system, 4
ceramic, vii, 18, 22, 25, 30, 31, 41, 44, 48, 51, 54, 55, 62, 63, 64, 65, 66
cheese, 1
cheesemaking industry, vii
chemical interaction, 19
chemical properties, 18, 21
Chromatographic Methods, 17
chromatography, 17, 18, 25, 61, 64, 66, 67
chronic active hepatitis, 15
circulation, 32, 45
clarity, 8
cleaning, 32
colon cancer, 15
colostrum, 18, 64, 67
Component velocity, 17
composition, 1, 29, 64
compounds, 6, 12
conditioning, 31
configuration, 30, 31
configurations, 31
conjugation, 65
cooling, 27
cortex, 4
cortisol, 65
cost, 1, 20, 21
covalent bond, 10
crystallization, 2
crystals, 2
cytochrome, 15

D

defence, 67
deficit, 6
dehydration, 2

denaturation, 8, 20, 21, 45
Denatured proteins, 19
deposition, 23, 38, 57
derivatives, 64
destruction, 11
Diafiltration mode, 21
diet, 61
differential scanning, 61
differential scanning calorimetry, 61
diffusion, 10, 18, 21
digestion, 3, 7
direct action, 14
displacement, 29
dissociation, 10
distillation, 3
distilled water, 32
dominance, 38
dressings, 62
drying, 2

E

egg, 12
electric charge, 17
electric field, 24
emulsions, 10
endotoxins, 14
environmental conditions, 19
environmental pollution, 1
enzymes, 4, 15
epithelial cells, 15
equilibrium, 45
equipment, 2, 20
esophagus, 15
ethanol, 3
evaporation, 2, 3
exclusion, 18, 21
extraction, 24

F

fat, 1, 3, 27
fatty acids, 6, 7, 8, 10, 61

Index

fermentation, 2
ferric ion, 13
fiber, 18
filtration, vii, viii, 2, 17, 18, 19, 20, 21, 22, 23, 32, 55, 62, 65, 66, 67
Filtration by membrane technology, 17
flexibility, 8
flocculation, 2
fluidized bed, 2
foams, 19
food industry, vii, 1, 3, 4, 10
formula, vii, 2, 4, 6, 15
fouling, 20, 21, 22, 23, 32, 37, 38, 57, 62, 65, 66
free radicals, 12, 14
freezing, 63
FTIR, 65
FTIR spectroscopy, 65
fungi, 14

G

gastrointestinal tract, 4
gel, 8, 17, 18, 66
gel formation, 8
gelation, 8, 10
glucose, 4
glycine, 7
glycomacropeptide fraction, 3
glycoproteins, 4, 11

H

hepatitis, 14, 15
herpes, 14
herpes simplex, 14
High Performance Tangential Flow Filtration, 21
host, 67
human immunodeficiency virus, 14
human milk, 6
hydrolysis, 4
hydroxyapatite, 18, 66

I

ideal, 18
image, 28
immune function, 64
immune response, 14
immune system, vii, 14
immunity, 12, 15
immunoglobulin, 11, 51, 55
immunoglobulins, viii, 11, 12, 18, 19, 65, 66
induction, 15
inflammation, 13, 14
inflammatory responses, 64
infrared spectroscopy, 61
inhibition, 15
inhibitor, 4, 14
initiation, 15
interface, 10
interferon, 15
intrinsic viscosity, 10
ion-exchange, 25, 61
ionic strength, 8, 10, 19, 21, 22, 23, 24, 25, 66, 67
ions, 22, 45
iron, vii, 12, 13, 14, 24
isolation, 15, 17, 18
isoleucine, 6

K

kinetics, 65

L

lactation, 4
lactic acid, 1
lactoferrin, vii, viii, 4, 12, 13, 14, 15, 23, 26, 32, 35, 38, 45, 54, 55, 57, 61, 63, 64, 65, 66, 67
lactose, vii, 1, 2, 4
lactose powders, vii
lakes, 1

leucine, 6
ligand, 18, 22
light scattering, 65
lipids, 7
liquid chromatography, 17, 18, 32, 63, 66
liquid phase, 19
low temperatures, 20
lysozyme, 10, 14, 22, 64

M

macrophages, 11
majority, 1, 18, 21, 24
manufacture, 1
manufacturing, 3
matrix, 18
membrane filtration, vii, 62, 65, 67
membrane permeability, 14
membrane technology, vii, 3, 24
membranes, 23, 24, 25, 30, 31, 63, 64, 65, 67
metabolism, 7, 13, 64
metabolites, 9
metastasis, 15
meter, 30
mineral salts, vii, 1
modification, 21, 22, 25
molar ratios, 45
molecular weight, 5, 7, 11, 12, 44, 57
molecules, 5, 8, 10, 13, 14, 17, 18, 22, 38, 44, 57
monomers, 8
mucosa, 15

N

neonates, 7
neurotransmission, 11
nitric oxide, 11
nitrogen, 6
NK cells, 11
nutraceutical, 5

O

opportunities, 67
optimization, 22
oral cavity, 12
organic matter, 1
organic solvents, 19
osmosis, 3
oxidation, 12
oxygen, 9, 11

P

particle size distribution, 2
pathogenic organisms, vii
pathogens, 14
peptides, 3, 63, 66
performance, 17, 18, 21, 32, 55, 63, 66, 67
peripheral blood, 4
pH, vii, 3, 5, 8, 9, 10, 19, 21, 22, 23, 24, 25, 26, 27, 30, 31, 32, 35, 36, 37, 38, 39, 41, 44, 45, 48, 51, 54, 55, 57, 62, 66, 67
pharmaceutical industry, vii, 2
phenylalanine, 6
physiology, 61
polarization, 20, 22, 32
pollution, vii, 1, 3
Polyanionic, 17
polycationic resins, 17
polypeptide, 5, 12
polystyrene, 63, 66
precipitation, 17, 19, 20, 25, 65
pressure gauge, 29
prevention, 67
probe, 29
production costs, 2
proliferation, 14, 15
promoter, 14
Protein fractionation, 21
proteinase, 65
protein-protein interactions, 20, 21, 22

Index

proteins, vii, 1, 3, 4, 6, 8, 10, 11, 12, 15, 17, 18, 19, 20, 21, 22, 23, 25, 35, 36, 38, 44, 45, 54, 55, 57, 62, 63, 64, 65, 66
proteolysis, 14
purification, vii, 2, 18, 19, 20, 21, 23, 25, 26, 63, 66, 67
purify lactose, 2
purity, 2, 18, 19, 20, 23, 24, 25, 26, 35, 55, 57

R

radicals, 9, 11
reactions, 10, 12
recommendations, iv
regenerated cellulose, 23
regeneration, 32
rejection, 22
resins, 17
resistance, 22, 31
resolution, 18, 21, 65
retinol, 7, 64
rights, iv
room temperature, 10, 32

S

salts, vii, 1
saturation, 12
Selective precipitation, 17, 20
selectivity, 20, 23
serotonin, 5, 64
serum, viii, 3, 4, 8, 9, 10, 11, 48, 55, 61, 63, 65, 66
serum albumin, viii, 3, 4, 8, 9, 10, 48, 55, 61, 63, 65, 66
shape, 11
shear, 45
shock, 14
small intestine, 15
sodium, 32, 66
solubility, 3, 8, 9, 17

solvents, 19
species, 4, 21
stabilizers, 62
steel, 30
storage, 2
streams, 29
sucrose, 2
suppression, 15
surface properties, 21
surface tension, 10
surfactant, 19, 63
suspensions, 63
synthesis, 4

T

tangential flow filtration, 20, 22, 23, 67
temperature, 3, 8, 10, 19, 23, 26, 27, 29, 31, 32
titanium, 30
transferrin, 12
Transmembrane pressure, 21, 32
transmission, viii, 22, 23, 24, 25, 26, 35, 37, 39
transport, 6, 38, 45, 66
trifluoroacetic acid, 32
trypsin, 4
tryptophan, vii, 5, 6, 64
tumours, 15
tyrosine, 6

U

ultrafiltration of whey, 3
ultrafiltration stage, 25, 26
underlying mechanisms, 67
unions, 13, 18

V

vacuum, 32
valine, 6, 7
velocity, 17, 23, 26, 32

viruses, 14
viscosity, 3, 9

W

waste, vii, 1, 32
waste treatment, 1

Whey, v, vii, 1, 3, 17, 27, 35, 65
whey protein concentrates (WPC), 3
whey protein isolates (WPI), 3

Z

zirconium, 22, 30